MW00477724

Pigs, Profits, and Rural Communities

SUNY Series in Anthropological Studies of Contemporary Issues
Jack R. Rollwagen, Editor

Pigs, Profits, and Rural Communities

Kendall M. Thu
and
E. Paul Durrenberger,
Editors

State University of New York Press

Published by
State University of New York Press, Albany

© 1998 State University of New York

All rights reserved

Printed in the United States of America

No part of this book may be used or reproduced
in any manner whatsoever without written permission.
No part of this book may be stored in a retrieval system
or transmitted in any form or by any means including
electronic, electrostatic, magnetic tape, mechanical,
photocopying, recording, or otherwise without the
prior permission in writing of the publisher.

For information, address State University of New York
Press, State University Plaza, Albany, N.Y., 12246

Production by E. Moore
Marketing by Nancy Farrell

Library of Congress Cataloging-in-Publication Data

Pigs, profits, and rural communities / Kendall M. Thu and E. Paul Durrenberger.
p. cm. — (SUNY series in anthropological studies of contemporary issues)
Includes bibliographical references (p.) and index.
ISBN 0-7914-3887-2 (hc : acid-free). —ISBN 0-7914-3888-0 (pb : acid-free)
1. Swine—United States. 2. Pork industry and trade—Environmental aspects—United States. 3. United States—Rural conditions. I. Thu, Kendall M., 1960– . II. Durrenberger, E. Paul, 1943– . III. Series.
SF395.8.A1P54 1998
306.3'49—dc21 97-35434
CIP

10 9 8 7 6 5 4 3 2

Contents

Acknowledgments

The crafting of a multiple-authored volume is a challenge requiring patience and flexibility among contributors. We thank all of the authors for their willingness to work together and for their unfailing responsiveness to our editorial requests and deadlines. A special note of gratitude is owed to Professor Walter Goldschmidt for his support of this effort and for reviewing all of the chapters. We hope this volume is a reminder of the enduring value of his theoretical and empirical contributions to understanding the role of food production in human adaptation.

Julia Venzke played a pivotal role in helping to edit chapters, prepare illustrations and tables, check bibliographies, and electronically assemble the volume. Without her keen assistance and sharp eye we would still be fumbling over our electronic ineptitudes. Thanks to Judy Thirtyacre for her assistance in preparing the details of this volume, and thanks also to W. Jamie Ward for his help in preparing the tables.

And finally we owe a debt of gratitude to the hundreds of farmers and other rural residents who invited us into their homes and neighborhoods to explain to us the changes occuring in their communities. We hope this volume provides others with a broader understanding of those changes and their significance for the health of our society.

Introduction

Kendall M. Thu and E. Paul Durrenberger

Across Iowa and the Midwest, from North Carolina and Pennsylvania in the East to Utah and Wyoming in the West, there are highly publicized debates, hearings, conferences, and trials concerning the environmental, social, and economic consequences of industrialized swine production. These events have captured national attention via a report by "60 Minutes", a Pulitzer prize winning newspaper series from North Carolina, the *Wallstreet Journal,* the *L.A. Times,* national network news coverage, National Public Radio, and from stories in national news magazines.

Much of this attention focuses on environmental problems, the displacement of family farms by nonfamily corporations, and rural communities organizing to oppose industrial swine operations. The debate is familiar. On the one hand are visions of self-contained, perpetually sustainable family farms representing all that is best in the ideals of the yeoman farmer of Jeffersonian America and the rural social and economic prosperity that accompanies widespread, small owner-operated farms. On the other hand are conceptions of highly efficient industrial food production, the prosperity industrialization brings to rural communities, and the cheap food it offers consumers.

There are many more than two positions on these issues, and it is the politicization of rhetoric that forces people into a dualistic way of thinking. Since legislative, administrative, and judicial processes are supposed to be governed by scientific findings, proponents for each position search for relevant scientific results upon which to base their claims, hence science becomes politicized (Durrenberger 1992; Thu 1996). Both the Jeffersonian and the industrial perspectives have respectable academic support, and each has vocal critics.

Industrialized agriculture is a form of food production heavily dependent on fossil fuel–based energy inputs such as fertilizers, machinery, pesticides, and gasoline. Based on measures of energy expended per calorie of food produced, industrial agriculture is the most inefficient form of food production in the history and prehistory of humankind. Industrial agriculture is characterized by large-scale operations that substitute capital-intensive fossil fuel–based technology for people. Capital-intensive technologies and inputs are increasingly sold to industrial farming operations by firms operating off the farm with expectations of profit and capital accumulation that diverge from the interests of farmers. At its most extreme, operational management prescriptions are handed down to farmers directing them in the amounts and types of inputs to use as part of doing business with off-farm corporations. In addition, ownership becomes separated from the community so that profits are externally defined and assigned with a purely economic denominator while local benefits and costs that include quality of life, the environment, and human values, such as mutual trust and sharing, are largely ignored. Concentration of production and distribution into relatively few firms which integrate highly specialized operations such as food processing and marketing is common. And a highly specialized political apparatus extending into national, regional, and local governing bodies, as well as state-supported institutions to support and replicate the system, is typical of human organizations associated with industrial agriculture.

In 1940 and 1941 Walter Goldschmidt conducted research on industrialized agriculture in California and its relationship to quality of life. The results were unequivocal. The emergence of an industrialized form of agriculture resulted in measurable declines in social and economic conditions and the appearance of an urbanized configuration of human relationships and adaptation. Goldschmidt's detailed research on two rural communities revealed better conditions in the community associated with farms where the majority were owner-operated and managed and labor was primarily family-based compared to the community surrounded by industrialized farms in which ownership, management, and labor were largely divorced from each other. His research has been replicated and substantiated at different times and locations over the past fifty years (MacCannell 1988; Durrenberger and Thu 1996).

In the face of Goldschmidt's findings and a substantial body of subsequent research, the industrialization of agriculture continues largely unabated. This raises the vexing question—why? The predominant response, anchored in the spirit of Adam Smith and the industrial paradigm that John Ikerd discusses in his chapter in this book, argues for the inevitability of economic forces to shape forms of production and consumption in a free market. This widely held view, led by economists and agribusiness leaders (Tweeten 1983; Urban 1991) asserts that farmers make necessary rational decisions to industrialize based on competitive market conditions in order to make profits and survive. Those holding this view respond to the problems raised by researchers in the Goldschmidt tradition by variously ignoring them, casting them as part of a scientifically disguised liberal political agenda, or using economic advantages they impute to industrialized agriculture as "evidence" to refute them. Frequently cited as evidence for the economic advantages of industrial agriculture are the wide availability of abundant food at extremely low cost to consumers and the record productivity and efficiency of modern farmers. In general, these economic advantages are viewed as outweighing the costs cited by researchers in the Goldschmidt tradition. Consequently, the problems cited by Goldschmidt and his contemporaries are hurdles to overcome in order to continue industrializing rather than signals of the fallibility of industrial agriculture.

One seemingly sensible approach to assessing industrial agriculture is to articulate and assess its advantages and disadvantages, a kind of cost-benefit analysis. For example, does the fact that U.S. consumers have cheap food outweigh problems created by industrialized forms of food production? As reasonable as it sounds, this approach ignores a fundamental component of industrial agriculture. An intrinsic feature of industrialized agriculture is the concentration of power over the production and distribution of precious resources as Senator Robert Morgan shows in his chapter on the political connections behind North Carolina's corporately dominated hog industry. Industrial agriculture has survived and prospered for the last half century because certain sectors benefit. Beneficiaries of industrial agriculture have a vested interest in predisposing the questions and rhetoric to assess its efficacy (Hightower 1973; Thu 1996). If profit is the principal measure of success, the more profit that is reaped, the better able the beneficiaries are to promote the advantages of industrialized agriculture and discredit even those who hold valid scientifically based countervailing views. This means that those in privileged positions can produce the knowledge necessary to support their vested interests.

Even when overwhelming scientific evidence attests to problems from the system that produces and distributes a particular commodity, such as tobacco, the enormous political and economic power attained by those in control counteracts rational assessment. Hence, the issue becomes not just one of producing and weighing objective evidence, but identifying those who are producing the evidence, how they are asking the questions, why they are producing the research, how it is being used, and who the beneficiaries are. As many of the chapters show in this book, the interlinkages between science and politics are intrinsic to industrialized food production and inherently tilt any assessment in its favor. Consequently, assessing the advantages and disadvantages of industrial agriculture is not simply a matter of examining the weight of scientific evidence, but also of examining biases in assumptions, the "paradigms" as John Ikerd calls them in his chapter, that underlie the evidence. For example, what are the assumptions that underlie economic measurements of productivity and efficient food production?

This volume focuses on the rapid industrialization of swine production in the United States and the consequences for its rural health, including the physical, social, psychological, economic, and political well-being of people. Most of these issues are not incorporated into economic analyses of food production, but as this volume demonstrates each has a distinct, often measurable, and very human cost. Contributors come from broadly contrasting disciplinary and vocational backgrounds, including a veterinarian, two farmers, a farm organization leader, a former U.S. senator, an agricultural economist, a medical psychologist, and anthropologists. Despite divergent backgrounds, a consistent theme emerges consisting of the broad rural health consequences of industrial swine production. Goldschmidt's findings that industrial agriculture results in a number of socioeconomic problems is as cogent today as it was in the 1940s. However, many of the issues raised in this collection go beyond those identified by Goldschmidt. Nonetheless, the underlying assumptions concerning the efficacy of industrial agriculture and many of its consequences remain remarkably similar.

This volume provides a broad and coherent picture of issues and problems associated with the industrialization of swine production. There is a reasonably well articulated dominant view of the advantages of industrialized swine production in the public domain. It is especially evident in the doctrines of various agricultural organizations and commodity groups. This volume identifies and challenges the assumptions of those views with concrete examples and evidence. As such, we can predict with reasonable certainty on the basis of Goldschmidt's experience (1978) that it will be castigated by proponents of industrial agriculture as an example of rhetoric from "antigrowth interests." In contrast, members of anti-industrial agricultural groups may embrace it as ammunition for their political causes. These typical responses reflect the point that for too many, the lines have been drawn, positions staked out, and interests entrenched, and for them this volume will do little beyond affirming our expectations or theirs. We can do no more than point out that we are not political activists and that we neither promote nor defend industrial agriculture based on any political agenda. Rather, as researchers with cross-cultural experience studying food production systems, we identify, analyze, and assess patterns of food production and their relationship to human adaptations.

There are many rural community members, policy makers, farmers, rural researchers, and students of agriculture who are honestly uncertain about the best direction. This volume is for them. The contributions presented here provide a legitimate counterweight to overly rosy rural development scenarios often painted for rural communities considering industrialized swine facilities. Moreover, it provides those who propose to promote, build, or operate such facilities a clear view of legitimate issues and problems they should consider. For still others, such as anthropologists, rural sociologists, and others interested in structural changes in agriculture, broad rural health issues, and their relationships to human adaptation, it provides theoretical and practical insight into a range of intertwined issues.

THE INDUSTRIALIZATION OF THE SWINE INDUSTRY

There are few topics, perhaps none, that take up more print and library space in the U.S. than agriculture. Except for air and water, nothing is more basic to human survival than food. The amount of research and printed materials on agriculture appears to have grown inversely to the declining number of farmers over the past sixty years. As a Norwegian farmer once quipped, "we keep getting more agricultural experts and more advice when there are fewer and fewer of us" (Thu 1996). Part of the reason for this is obvious, researchers and various agricultural experts are among the beneficiaries of an industrialized form of agriculture. A cycle of dependency emerges in which new methods of food production that require technical expertise are offered only to be discarded as obsolete as new innovations are recommended to farmers to enhance their competitiveness.

The industrialization of swine production fits a well-documented pattern of technological innovation in agriculture. As Jim Braun, himself a pork producer, describes in his chapter, the most evident of these innovations is the development of confined swine production technologies in the early 1970s that moved hogs from pastures and partial shelters to completely enclosed facilities specially designed to control each step of the production process. These long, low-lying metal buildings are arranged in multiple rows with no signs of farm houses, barns, farm implements, or even livestock traditionally apparent on family farms. In their place, open air waste pits called "lagoons", metal office buildings, and the frequent arrival and departure of large transport trucks provide an industrial factory appearance to pork production. A city dweller would be hard-pressed to figure out what sort of livestock is being produced, since many hogs in these facilities do not see the light of day except on the way to market. Many of these changes are patterned after technological developments in the poultry industry as John Morrison, executive director of the National Contract Poultry Growers Association, discusses in his chapter.

Factory-style swine production facilities are fairly standardized, consisting of multiple long, low-lying metal buildings set atop concrete foundations, rows of ventilation fans protruding from the sides and ends, elevated bulk feeder tanks with feed augers to move feed inside, and frequently a nearby outdoor lagoon to hold animal waste. The interiors are also similar, characterized

Figure 1
Family Hog Farm.

Figure 2
Industrial Hog Operation.

by rows of holding pens, slotted floors that allow waste to wash into a pit below where it is later pumped or flushed to an outdoor lagoon for anaerobic decomposition, auger-driven feeding pipes fitted along the ceiling with outlets to feeding troughs connected to each pen, automated watering systems, and walkways between rows of pens. Enclosed facilities allow better control over factors critical to growth and profits, including exposure to climatic and other environmental fluctuations (particularly in temperate climates), feeding regimens, and reproduction. In addition, genetic selection provides swine that convert feed to meat more quickly and with less fat, so leaner hogs are market ready sooner. Differential weaning processes based on sex, separate and frequently quarantined production facilities for each stage of growth, and the use of antibiotics all contribute to increased disease control. In addition, capital investments in technological advances reduce the amount of labor required to produce each hog.

These technological changes accompany rapid structural changes. The number of swine producers has declined precipitously from 750,000 in 1974 to 157,000 by the end of 1996 (USDA). Between 1994 and 1996 one out of every four hog producers left the business. And by the end of 1996, 3 percent of producers representing the largest corporate operations in the United States produced 51 percent of all the hogs, while 21 percent of all hogs were grown under contract—a situation fraught with social and economic problems that John Morrison describes in his chapter on the poultry industry. Producers that remain are larger and increasingly use outside capital to purchase new technologies. The traditional swine producing areas of the Midwest and the Great Plains are facing increasing competition from nontraditional areas such as North Carolina, Colorado, and Utah utilizing the latest technology and cheap labor to offset the cost of grain imported from the Midwest for feed. As a consequence, locally grown or purchased inputs for the production process are replaced with increasingly larger shares of purchased inputs from outside the locality (Chism and Levins 1994; Lawrence 1994).

Ownership, management, and labor become separated both economically and socially. In one form, producers contract with large companies or corporations to produce hogs. In these "risk-sharing" ventures, the farmer produces hogs owned by someone else following a standardized production protocol specified by the contractor. In such a system, wealth and profits are extracted from local economic systems which, even where standards of living increase, become increasingly dependent on control from outside systems. Yet total production per farm increases, with a concomitant increase in production efficiency as measured by output per unit of labor. However, profit margins narrow as increasing output saturates markets, pushes consumer prices lower, and pits farmers against farmers in a competitive "treadmill" of producing more product for less money (Durrenberger and Thu 1996). This is particularly true for farmers who, in contrast to food processors and distributors, are not in a position to pass along increased production costs. In contrast, vertically integrated systems wield complete control over the entire production and distribution process, from genetic stock to the grocery store shelf, so corporations can track and adjust costs across the entire food production and distribution process to maximize profits.

Consumer demand for pork in the United States has remained steady for the past ten years (*Feedstuffs* 1994). However, pork's share of total domestic meat consumption has declined since 1950 (Van Arsdall and Gilliam 1979). Nonetheless, pork accounts for 40 percent of world meat consumption, the single largest source of animal protein (United Nations 1991). U.S. pork exports have soared in the past couple of years, outstripping imports for the first time. Many of these burgeoning markets are appearing in eastern Asia, particularly Japan and the Pacific Rim. Markets in Mexico and Canada continue to be strong as well. Many of the larger swine producers are tailoring their meat to the specific consumer preferences of these markets, for example marbled pork in Japan. Moreover, some U.S. consumers are expressing an increasing desire for leaner meats of all types, including pork.

Swine industry leaders (Keppy 1995), agricultural economists (DiPietre and Watson 1994), and others (Williams and Pfouts 1995) point to consumer demand as a driving force behind industry changes. It is true that some consumers desire leaner meat, but it is not accurate to say they are therefore demanding industrialized production facilities. If industrialized swine facilities were the only form of production capable of producing leaner meat, such an argument might have merit. Randy Ziegenhorn describes one such alternative in his chapter on independent swine producer networks. There are other alternatives as well, including swine hoop structures resembling tent-covered wagons that are beginning to appear on the Iowa landscape. These facilities are far less capital intensive, more acceptable to neighbors and rural communities, and can be quite profitable for independent producers (Honeyman 1994).

It is true that meat packers are demanding leaner, more standardized carcasses and that larger producers tend to meet these demands better. Similar to other food industries such as fishing (Durrenberger 1996), it is the signals from processors, not consumers, that producers receive and act on. However, the profits of packers, processors, and retailers mitigate the signals of consumers when making demands upon swine producers. As Mark Grey describes in his chapter on the packing industry, a standardized hog carcass allows for the use of cheap, unskilled labor in a factory-style disassembly and packing process. The importation of foreign labor has a range of community costs that Grey details in a rural Iowa community. Is a processor's demand for a standardized carcass the same thing as consumer demand, or are packers manipulating the rhetoric of consumer demand to justify their demand for standardized carcasses in order to import unskilled labor from Mexico and Southeast Asia? This raises a more fundamental question of distinguishing the rhetorics of profits and consumer preferences in shaping changes in the organization and technologies of food production.

Consumer preferences are influenced by a number of complex factors (Griffith and Johnson 1989), involving much more than a simple single-dimensional desire for something such as "leanness." For the past fifty years, the real trend in consumer demand has been for food requiring less preparation time and effort. In the 1960s, approximately two-thirds of all consumed food came from grocery stores and only one-third from food service operations (Wolt 1996). Thirty years later consumers get more of their food from food services than from grocery stores. As Christopher Wolt, director of Strategic Food Resources for Noble and Associates, pointed out at the 1996 National Forum for Agriculture, health considerations such as cholesterol, fat, and salt content are becoming less important factors in consumer food choices. Approximately eight percent of the U.S. population, actually makes primary food consumption decisions based on health-related factors such as leanness. If changes in the pork industry were truly a response to consumer preferences, we would be seeing not an increasingly lean pork, but pork that is easier for the consumer to prepare.

Some consumers do want leaner meat, but their demand has little to do with how that lean meat is produced. The demands of industries that provide the building materials, veterinary supplies, feed supplements, and other inputs for this new type of swine production are likely more powerful than consumers in shaping forms of food production. Consumers buy what they are offered. Furthermore, there is no reason to assume that consumer preference for a particular food item is the same thing as a demand for a particular technology or form of production to produce that food item.

Even if consumer demand has little role in determining forms of swine production, conventional economic wisdom of the industrial paradigm suggests that large-scale industrial swine production facilities must be productive, efficient, and most of all profitable, otherwise they would not be proliferating and surviving while vast numbers of smaller, independent operations are disappearing. And if this is the case, is it not reasonable to assume that a more efficient form of production is better for the overall rural health of agricultural regions in the United States? Much of this line of reasoning is predicated on assumptions rooted in economies of scale, as well as assumptions about which factors should be included and considered in measurements of productivity, efficiency, and profitability. Assumptions are not facts.

Economies of scale refer to situations in which a constant proportion of resource inputs results in a disproportionately higher rate of outputs or profits simply because of the size of the operation or business. For example, a farm with one thousand hogs creates more than twice the output and/or net profits than does a five hundred-hog operation. Similar advantages might be gained even if resource inputs are not proportionately increased with growth, as long as net profits increase faster than an increase in inputs. For example, ten hogs might be added to an existing operation with two hundred head without any increase in heating costs, relatively small increases in water usage and labor, and proportionately modest increases in maintenance and veterinary costs, which make the increased cost of feed worthwhile because of proportionately increased profits.

Some argue that the rapid growth of large-scale swine operations is occurring because their scale allows them to achieve increases in production efficiency and economic profitability by virtue of their size. However, we are aware of no data on the production efficiencies and profitability of the largest swine operations in the country, particularly the largest fifty seven which market more than fifty thousand hogs each. This data is closely guarded proprietary information of the firms themselves. Data do exist on the production efficiencies and profitability of most swine producers. Studies (Iowa Farm Business Association 1992; Mueller 1993) indicate economies of scale are achieved in very modest-sized swine operations, and that size has little to do with efficiency of production and profitability. In an analysis of 705 hog farms in Illinois, Mueller (1993:4) showed that critical factors for profitability were the number of pigs weaned per litter, weight gain efficiencies, price of feed, and market price. In a step-wise regression analysis, size of operations explained less than 5 percent of the variation in profitability. In other words, scale was the least important determinant of profitability. These findings substantiate earlier studies conducted on agricultural operations generally which show that modest-sized operations capture the bulk of economic efficiencies, and larger operations have little to no advantage in terms of economies of scale (see Madden 1967 for a summary).

The fact that large-scale operations are not inherently more efficient than their smaller counterparts does not mean smaller operations are efficient. What it does mean is that smaller operations are not inefficient by virtue of their size. Analyses of swine production records shows a range of production efficiencies; some producers are more effective than others. The response of national and state pork producer groups, farm organizations, and many agricultural economists is to place industry problems on the shoulders of independent producers whom they cite as poor managers and inefficient producers. For example, listen to the chastising editorial words of the editor of *Pork 94* (Miller in *Pork 94,* 5 October) a monthly magazine for pork producers: "I can't help wonder what the payoff would be if some producers put as much time and energy into upgrading their own operation as they spend worrying about the 'big boys.' The real question should be: What have you done for yourself lately? It doesn't matter how much money the other guy makes. What's important is that you're profitable. If you're not, why not?"

The dominant rhetoric of swine industry change focuses on problems of production efficiency among individual producers. Producers are cited for poor record keeping, use of substandard genetics, inadequate herd management, and their reluctance to expand and invest in the technological tide of the future. This approach assumes that efficiency is the key to survival. However, if larger producers are not inherently more efficient than their modest-sized counterparts, why do they continue to survive while smaller producers are disappearing? A few facts cast light on an important dimension of this trend and challenge assumptions about efficiency.

Approximately thirteen percent of U.S. pork production comes from the fifty-seven largest producers; each producing more than fifty thousand hogs annually. These fifty-seven companies have increased production at a rate that far exceeds that of smaller producers (Grimes and Rhodes 1994), despite the fact that domestic consumer demand has remained constant. In 1994, this resulted in a deluge of hogs into the market that set national production records. As a result, hog prices nose-dived from fifty cents to less than twenty-seven cents a pound. In real dollars, this is the lowest price since the Depression. For most family farm producers, cost of production is approximately thirty to forty cents a pound. How long can even the most efficient family hog producer survive losing ten cents on every pound of hogs? Large-scale corporate producers gobbled up domestic market share at the expense of independent family producers if for no other reason than their massive production. Large-scale producers were not just responding to open market pressures of supply and demand, they were creating them. The issue was not who was most efficient, rather who had the resources to shape the structure of the market. Furthermore, as producers and processors collaborate, there simply is no market.

A comparable point was made concerning a prevailing misguided view that farms that were weeded out during the 1980s farm crisis were inefficient. As the agricultural economist Neil Harl demonstrated, efficiency was largely irrelevant to survival during the 1980s crisis. "What occurred in agriculture in the 1980s in terms of firms failing because equity was exhausted or operating credit was denied had little to do with efficiency and did not represent a continuation of the long-term trend toward greater efficiency in agriculture. In fact, the firms at risk were some of the most efficient in the industry and were operating at or near the point of greatest efficiency" (Harl 1990: 20). The idea that changes in the swine industry reflect a natural process of survival of the most efficient is also an assumption, not a fact. It may well be that, similar to the 1980s, survivors are those who have access to the most capital by establishing appropriate "relationships" with economic power centers, as the pork producer

Jim Braun discusses in his chapter. If this is the case, adaptive responses by individual producers to become more efficient by focusing on reducing their costs of production could be misguided.

Even more fundamental is a critical examination of assumptions concerning how efficiency is conceptualized and measured in agriculture. Efficiency of production is typically measured by examining the cost and amount of inputs and related expenses compared to outputs consisting of food commodities and profit. The operation that can produce the most food with the fewest and least costly inputs at a maximum profit is the most efficient. Types of inputs and related expenses usually included in measures of efficiency and productivity are things such as land, machinery, fuel and oil, utilities, labor, seed, feed, veterinary supplies and services, fertilizer, herbicides, pesticides, insurance, building repair, machinery and building depreciation, and interest, among others (Iowa State University 1991). Outputs include crops and livestock produced. Profits are calculated by subtracting the cost of inputs and related expenses from the money received from marketing crops and livestock. Efficiency is then calculated by comparing various dimensions of the relationships between inputs and outputs. For example, machinery costs per acre, cost of livestock feed per unit of animal growth, expenses per bushel of grain, and more general calculations such as value of farm production per dollar of expense or percentage return relative to total equity. But certain measures are utilized to calculate efficiencies while others are excluded. For example, farming is among the most dangerous occupations in the country, with known elevated rates of disabling illnesses such as respiratory diseases among swine confinement production workers described by Kelley Donham in his chapter in this volume. What are the productivity and health care costs of respiratory health problems among the 25 to 30 percent of swine confinement production workers known to be experiencing problems? Conversely, what are the educational, community, social, and human character benefits of learning honesty, hard work, ingenuity, flexibility, and fairness as part of being reared in a farm environment? Why are these beneficial factors not included in calculations of efficiency? To respond that they are intangible, immeasurable, and therefore less relevant is a manifestation of a pervasive and powerful assumption of what economic efficiency and productivity are. Again, these are assumptions, not facts. As DeLind and Nickles show in their chapters, the social and community costs of corporate swine facilities are all too real. To be sure, farmers must make a profit if they are to survive. No one pays a tractor repair bill with strong character. But rural communities, policy makers, and society as a whole must understand, judge, and assess the total picture of farming and its relationship to our rural and national health.

Understanding large-scale industrial swine production facilities includes examining not only their internal operational efficiencies but also their relationship to rural communities in which they are situated and society as a whole. A number of rural development reports have been issued enthusiastically endorsing the economic benefits of large-scale and contract swine production (DiPietre 1992; DiPietre and Watson 1994; Thornsbury, Kambhampaty, and Kenyon 1993). Glowing projections concerning job creation, increases in personal income, increase in tax revenue, business growth, and general increases in revenue are typical of these reports. However, the projected benefits are based on a set of assumptions, not empirical data, and factors considered in the projections are uniformly biased in favor of benefits. In fact, the economic model (IMPLAN) used for these reports was originally developed by the U.S. Forest Service and is virtually incapable of calculating negative impacts or costs. Because of this weakness, costs are relegated to a residual category at the end of such reports, and it is asserted without evidence that these costs will not be incurred. An example comes from the end of a 1994 University of Missouri Commercial Agriculture Program report on large corporate swine production in Missouri. "In addition, some people fear an environmental disaster may occur from a major accident in waste handling methods. Such an accident might pollute both surface and ground water and possibly harm wildlife, especially fish in streams or waterways affected [sic]. The probability of a major negative environmental impact beyond odor is unlikely" (DiPietre and Watson 1994: 52).

The assumption that an environmental impact is unlikely was proven false in the summer of 1995 when the large corporate swine producer described in the University of Missouri report was caught in three major manure spills killing over 178,000 fish in public waters according to the Missouri Department of Natural Resources (November, *Pork 95:* 67–68). In the fall of the same year, another corporate hog producer in northern Missouri was fined for a manure spill that traveled over eight miles down a tributary of the Grand River killing nearly ninty thousand fish (*Pork 95:* 68). These were part of a series of eight separate manure waste spills into Missouri public waters resulting in environmental fines of $170,000. That same year, in north-central Iowa, site of the most rapid proliferation of large-scale hog production in that state, 1.5 million gallons of manure flowed from a hog lagoon into a tributary of the Iowa River (McMahon 1995: 17). And in North Carolina, the state with the most rapid expansion of corporate swine production in the country, 22 million gallons of swine feces and urine escaped an eight-acre waste holding lagoon, eventually draining into the New River where it destroyed most of the aquatic life in a seventeen-mile stretch (Satchell 1996). The mounting number of these environmental problems demonstrates that, contrary to economic development rhetoric and assumptions, they are very real costs.

As Laura Jackson details in her chapter on water quality and large-scale swine operations, our ability to understand and address environmental issues is limited by a myopic view of agricultural and environmental systems. In addition to detailing specific issues such as swine waste handling systems, field application of swine manure, and nitrogen deposition rates, Jackson reminds us that it is the total ecology and economy that includes a reliance on monocropping that provides hog feed that must be included to understand and address environmental issues. Similarly in his chapter, John Ikerd urges us to broaden our gaze in pointing out that a large-scale corporate swine facility may indeed have economic benefits for a circumscribed geographical area. But when considered as part of wider rural and swine industry health there are measurable disadvantages and costs. This is because a fundamental part of industrialized food production is replacing labor with capital-intensive technology and reducing the amount of labor required per unit of output. Consequently, it is inherently misleading to promote job growth benefits of industrial swine facilities since by the very nature of their "efficiencies" they are intended to displace labor. As Ikerd points out, employment gains in one area are more than offset by employment losses in others. In other words, the jobs of workers in swine production factories do not compensate for losses of independent owner-operators.

Ikerd's findings confirm for the swine industry what established research has shown for agriculture generally, that the emergence of industrial agriculture is a significant precipitating factor for the decline of rural social and economic conditions, particularly in areas dependent on agriculture (Barnes and Blevins 1992; Goldschmidt 1978; Lobao 1990). This research includes a number of regional and nationwide assessments examining various measures of industrialized agriculture, for example, scale, extent of hired labor, and its effects on various indices of social and economic health, such as, poverty rates and food stamp usage. Our study in Iowa similarly showed a measurable relationship between large-scale swine production and measures of declining economic well-being in rural Iowa (Durrenberger and Thu 1996). Our analysis indicated that having more hog farms was better for rural health than producing more hogs. Despite this evidence, the dominant criterion of industry health continues to be the absolute number of hogs produced and a state's proportion of national production.

In addition to traditional economic considerations such as employment, there are a number of other social and community costs. Susan Schiffman, an expert in odor research, demonstrates in her chapter that people living in the vicinity of large-scale swine facilities experience clear and measurable declines in psychological health. The gases and corresponding odors emitted from these facilities have been widely reported by people experiencing them as atrocious, repugnant, and unbearable. Subsequent research suggests neighbors may be experiencing physical health problems similar in kind, though less intense, than well-documented health problems among swine confinement workers (Thu et al. 1997). Unfortunately, swine-generated odor and gases and people's responses to them are all too frequently treated by officials and many researchers as a matter of individual perception and subjective experience (Thu and Durrenberger 1994). Considerable research funds are being channeled into developing odor control technologies. Strangely, the availability of research funds indicates a clear recognition of a problem, but when problems are experienced by neighbors they are deemed too elusive or subjective to address. The burden of proof then falls on the neighbors to demonstrate the reality of their problems. Schiffman's work clearly shows that these problems are real and that they have a physiological basis.

In her contribution, Laura DeLind provides perhaps the only relatively complete case study of the community consequences of a large-scale corporate swine facility. She traces an agonizing cycle from the arrival of a facility in a Michigan community, to the community's response to its eventual departure. In contrast to most cases, this Michigan community "successfully" ridded itself of the corporate swine facility. However, the human cost of the battle to purge itself of the facility exhausted the community to the extent that it was forever changed. DeLind describes in detail the human dimensions of this battle and how the human social costs that are excluded as intangible externalities in economic development models are in fact the all-too-vivid realities of actual people. These are facts, not assumptions.

As Iowa family farm hog producers for over twenty years, Jim Braun and his wife Pam discuss the myths and realities of industry change in their chapter. Increasingly surrounded by corporate hog operations in north-central Iowa where their hog farmer neighbors have all but disappeared, the Brauns face a daunting challenge in trying to forge a farming future for their children and resist the transformation of their small community into a company town. With the realization that industry changes are not simply the result of so-called natural market forces, Jim Braun saw the absurdity of industry rhetoric focusing on increasing operational efficiency as a survival strategy for his farm. No longer wedded to his father's admonition to stay on the farm and work hard, he is now working to keep his family hog operation by sowing and tending the political fields of the state legislature and farm organizations where he works relentlessly for an equal chance to survive.

Robert Morgan, a former U.S. senator and state attorney general from North Carolina, and Blaine Nickles and Jim Braun—farmers in north-central Iowa, describe the legal and political dimensions of industrial swine production in North Carolina and Iowa. As a farmer, Blaine Nickles describes his experiences with the consequences of large-scale swine production in his community. Odor, water quality, economic impact, and changing social relationships are among the problems he discusses. In addition, he traces his grassroots efforts to have the problems of his community redressed through political processes. A disconcerting picture emerges of what happens when industry power becomes concentrated, blocking political channels for getting the real problems of rural citizens addressed. From a different vantage point, Senator Morgan describes the political and legal dimensions of corporate swine production in North Carolina. North Carolina is home to the greatest concentration of corporate swine production in the United States, purposefully courting pork production to replace its troubled tobacco industry. As Senator Morgan points out, large-scale corporate hog producers wield political and legal power that puts typical rural residents at a distinct disadvantage when they seek redress for problems they experience. Both chapters illustrate how the concentration of political power that accompanies the growth of corporate swine production works against the principles of a democratic society. These costs are among the most disturbing and have perhaps the most fundamental consequences for the rural health of the United States.

John Morrison, executive director of the National Contract Poultry Growers Association, outlines disturbing parallels between current changes in the swine industry and those leading to the complete vertical integration of the poultry industry some thirty years ago. He discusses the inherent power imbalance between the handful of poultry contractors who now control production and the growers who must accept contracts with them if they want to stay in business. A number of legal battles have raged over corporate contractors' exploitation of individual growers. As Morrison points out, current state and national pork producer organizations have a clear historical record from the poultry industry from which they can base decisions for the future of the swine industry.

In contrast to the fatalistic rhetoric of economic inevitability, a basic point of this collection is that we do have choices over the type of food production we want in this country. If the industrialization of food production were simply a matter of free market economics, then why does agriculture and the food industry have so many public and private agencies and organizations spending so much time and money attempting to influence policies and the politics of food production? If the market were the only reality, none of these would be necessary. The fact that more of these organizations exist now than ever before in the face of record small numbers of food producers is a testament to whom the beneficiaries of industrialized agriculture are and belies the rhetoric of free market forces. As discussed earlier, industrial agriculture has its beneficiaries, and it goes without saying that those that benefit will attempt to protect their interests. The notion that economics is an isolated sphere of activity into which politics does not intrude is a myth. Almost all state governments intervene to regulate systems of food production (Clunies-Ross and Hildyard 1992; Friedmann and McMichael 1989; Lindert 1991; Michelmann, Stabler, and Storey 1990; Timmer 1991). Assumptions that define and promote ideas of productivity and efficiency as inevitable and natural economic processes resulting in industrialized swine production convinces those who suffer its costs that there is nothing they can do about it. While this rhetoric is repeated in small town coffee shops, cafes, and grain elevators, the largest agribusinesses and food processors in the world are busy in the corridors of Washington, D.C., contradicting their free market rhetoric merely by participating in political processes. The predictable response is that their participation in political processes is simply to ensure fair and equitable markets and conditions. We leave it to the reader to decide by casting aside assumptions and weighing the facts.

Recent strides in a "movement" known as sustainable agriculture have succeeded in establishing a foothold relative to the dominant industrial agricultural paradigm. Sustainable agriculture is often perceived to be associated with specific farm practices to deal with issues such as soil erosion or decreasing fertilizer use while maintaining profitability (Keeney 1990). It may also include features such farmer health and safety as Donham and Thu (1993) advocate. Criticism often is leveled at sustainable agriculture that it is nebulous and ill-defined, and that it has no pragmatic program or mechanism for ensuring profitability among farmers who make the transition. However, any newly emerging system of adaptation requires a range of variation to increase its likelihood of success. Sustainable agriculture would be doomed to failure if its tenets were clearly defined and rigidly practiced at the outset. The underlying point of sustainable agriculture is not precisely what it is, but what it means (Duffy 1994) and why it is emerging. It is emerging because of defects in the industrial paradigm of farming, and whatever its resultant forms, they should be developed in response to clearly identified problems inherent in the industrial paradigm. The chapters in this book help to crystallize problems associated with the industrial paradigm in order to remedy them.

In the volume's Conclusion, Walter Goldschmidt reminds us of the human meaning and cultural evolutionary significance of changes in the way we produce our food. As a leading anthropological scholar schooled and experienced in studying the ways of human adaptation, Goldschmidt points to the waves of consequential changes that occur when we fundamentally alter a basic infrastructure of society—our food production system. Goldschmidt reminds us that this is not an issue of technological reticence or advocating a return to a simpler romantic past. Rather, it is an issue of cultural evolutionary cognizance, of learning what changes in food production mean for the social, economic, and political health of a society and the lessons that we ought to be wise enough to learn and courageous enough to apply. It is an anthropological lesson born of understanding the ways of human existence and adaptation, a lesson that reminds us that the challenge of today's prairie frontier is taming the all-consuming economic dragon that lays waste the fertile soils of human dignity and community.

REFERENCES

Barnes, Donna, and Audie Blevins. 1992. Farm structure and the economic well-being of nonmetropolitan counties. *Rural Sociology* 57(3):333–346.

Chism, John W., and Richard A. Levins. 1994. Farm spending and local selling: How do they match up? *Minnesota Agricultural Economist* Spring 676:1–4.

Clunies-Ross, Tracy, and Nicholas Hildyard. 1992. The politics of industrial agriculture. *The Ecologist* 22(2):65–71.

DiPietre, Dennis. 1992. The economic impact of increased contract swine production in Missouri. Staff paper. Columbus, Mo: Department of Agricultural Economics.

DiPietre, Dennis, and Carl Watson. 1994. The economic effect of premium standard farms on Missouri. Columbia, Mo: University Extension Commercial Agriculture Program, The University of Missouri.

Donham, Kelley J., and Kendall M. Thu. 1993. Relationships of agricultural and economic policy to the health of farm families, livestock, and the environment. *Journal of the American Veterinary Medicine Society* 202(7):1084–1091.

Duffy, Mike. 1994. Duffy details current sustainable ag issues. *Leopold Letter* 6(4):1, 9–11.

Durrenberger, E. Paul. 1992. *It's all politics: South Alabama's seafood industry.* Champaign: University of Illinois Press.

———. 1996. *Gulf Coast soundings: People and policy and the Mississippi shrimp industry.* Lawrence: University Press of Kansas.

Durrenberger, E. Paul, and Kendall M. Thu. 1996. The expansion of large-scale hog farming in Iowa: The applicability of Goldschmidt's findings fifty years later. *Human Organization* 55(4):409–415.

Feedstuffs (Reference Issue). 1994. *A complete reference and resource guide for the feed industry,* vol. 66, no. 30. Miller Publishing Company.

Friedmann, Harriet, and Philip McMichael. 1989. Agriculture and the state system. *Sociologia Ruralis* 29(2):93–117.

Goldschmidt, Walter. 1978. *As you sow: Three studies in the social consequences of agribusiness.* Montclair, N.J.: Allanheld, Osmun & Co. Publishers.

Griffith, David C., and Jeffrey C. Johnson. 1989 "'Eat Mo' Fish:' Using Anthropology to Increase and Diversify U.S. Seafood." In *Marine Resource Utilization: A Conference on Social Science Issues,* ed. J. S. Thomas, et al., 173–176. Mobile: University of South Alabama Publication Services.

Grimes, Glenn, and James V. Rhodes. 1994. "Small Survey Indicates Large Producers Growing Rapidly," *Feedstuffs,* 6 June, 26–27.

Harl, Neil. 1990. *The farm debt crisis of the 1980s.* Ames: Iowa State University Press.

Hightower, Jim. 1973. *Hard tomatoes, hard times.* Cambridge, Mass.: Schenckman Publishing Co.

Honeyman, Mark. 1994. A sustainable model for swine production. *Leopold Letter* 6(3):1, 11.

Iowa Farm Business Association. 1992. Iowa Farm Business Association. Economic Costs of Hog Production by Number of Hogs Marketed.

Iowa State University. 1991. *Iowa farm costs and returns.* Ames: Iowa State University Extension.

Keeney, Dennis. 1990. Sustainable agriculture: Definition and concepts. *Journal of Production Agriculture* 3: 281–285.

Keppy, Glen L. 1995. Foreword. *Market access situation analysis.* Des Moines, Iowa: National Pork Producers Council.

Lawrence, John D. 1994. A profile of the Iowa pork industry, its producers, and implications for the future. Staff Paper 253. Department of Economics, Iowa State University.

Lindert, Peter H. 1991. "Historical Patterns of Agricultural Policy." In *Agriculture and the State,* ed. C. P. Timmer, 29–83. Ithaca, N.Y.: Cornell University Press.

Lobao, Linda M. 1990. *Locality and inequality: Farm and industry structure and socioeconomic conditions.* Albany: The State University of New York Press.

MacCannell, Dean. "Industrial Agriculture and Rural Community Degradation." In *Agriculture and Community Change in the U.S.: The Congressional Research Reports,* ed. L. E. Swanson, 15–75. Boulder, Colo: Westview Press, 1988.

Madden, J. Patrick. 1967. *Economics of size in farming: Theory, analytic procedures, and a review of selected studies.* Agricultural Economic Report No. 107. Washington, D.C.: USDA Economic Research Service.

McMahon, Karen. 1995. Averting a manure spill. *National Hog Farmer,* 15 September: 17–20.

Michelmann, Hans J., Jack C. Stabler, and Gary G. Storey. 1990. *The political economy of agricultural trade & policy: Toward a new order for Europe and North America.* Boulder, Colo.: Westview Press.

Mueller, Allan. 1993. Economies of size in hog production: Is size related to profitability? *Farm Economics.* 93–95, Department of Agricultural Economics, University of Illinois at Urbana-Champaign.

Pork 95. 1995. "Premium Standard Spills Spur Hog Farm Critics" 15 September:67–68.

Satchell, Michael. 1996. Hog heaven—and hell. *U.S. News & World Report,* January 22: 55–59.

Thornsbury, Suzanne, S. Murthy Kambhampaty, and David Kenyon. 1993. Economic impact of a swine complex in Southside Virginia. Department of Agricultural and Applied Economics, College of Agriculture and Life Sciences, Virginia Tech.

Thu, Kendall. 1996. Piggeries and politics: Rural development and Iowa's multibillion dollar swine industry. *Culture and Agriculture* 53: 19–23.

———. 1996. What's a "year's work" worth? The influence of the state on cultural constructs of farming *Human Organization*55(3):289–297.

Thu, Kendall M., and E. Paul Durrenberger. 1994. Industrial agricultural development: An anthropological review of Iowa's swine industry. Paper presented at the twentieth Annual National Association of Rural Mental Health Conference, 3 July, Des Moines, Iowa.

Thu, Kendall M., K. Donham, R. Ziegenhorn, et al. 1997. A control study of the physical and mental health of residents living near a large-scale swine operation. *Journal of Agricultural Safety and Health* 3(1):13–26.

Timmer, C. Peter. 1991. "The Role of the State in Agricultural Development." In *Agriculture and the State,* ed. C. P. Timmer, 1–28. Ithaca, N.Y.: Cornell University Press.

Tweeten, Luther. 1983. The economics of small farms. *Science* 219:1037–41.

United Nations. 1991. *Food and Agricultural Organization.* New York, N.Y.

Urban, Thomas N. 1991. Agricultural industrialization: It's inevitable. *Choices* fourth quarter: 4–6.

Van Arsdall, Roy N., and Henry C. Gilliam. 1979. "Pork." In *Another Revolution in U.S. Farming?* ed. Lyle P. Schertz, et al., 190–254. Washington, D.C.: USDA.

Williams, Mark, and Barry Pfouts. 1995. Distribution and sale to the ultimate user: The second half of the equation. *Market access situation analysis.* Des Moines, Iowa: National Pork Producers Council.

Wolt, Christopher. 1996. Consumer food: World of tomorrow. Presentation at the National Forum for Agriculture, Friend or Foe? Technology and the Structure of Agriculture. Des Moines, Iowa.

Part I

Rural Community Consequences

We often refer to technology as though it has a life and purpose of its own, as in the phrase "technology driven." We also define and orient periods of human history and prehistory by types of technology, such as the "stone age," the "industrial revolution," or "the information age." Similarly, "industrial agriculture" orients us to a period of human history which evokes technological images of food production that include substantial tools, such as the tractor or the combine. Indeed, whole research agendas, bodies of research funding, colleges of agriculture, and agricultural policies have been geared toward pursuing and promoting an industrial form of agriculture as defined by its technology. However, no technology has a life, history, or purpose of its own. Technology never exists apart from people.

As the three chapters in this section describe, large-scale industrial hog production technology has ramifications for how people live together, the quality of their lives, the ability of farm families to survive, the quality of education and health care, and even people's sense of justice and fair play. These chapters also demonstrate that people understand and relate to technology as an extension of the people who use it. The ownership, management, and laborers within large-scale swine production operations and meat processing plants are identified relative to an existing community social arrangement.

In the initial chapter, the anthropologist Laura B. DeLind describes how a small Michigan community responded to the encroachment of large-scale swine operations. She describes how the process that led to the building of these facilities violated core community values, how community members suffered from indifference and disrespect, and ultimately how the community lost even in the face of victory. It is a tragic story of the loss of human spirit and erosion of faith in our democratic system.

From inside the swine industry, Jim and Pam Braun, themselves hog producers for over twenty years, detail the history of their family farm, their community, and the disruption caused by the spread of large-scale hog operations near their farm and community in northern Iowa. They reveal how they came to realize from firsthand experience that the industrialization process was not about economic progress or enhancing production efficiency, but rather about political and economic power and about corporate winners and community losers. They demystify industrialized swine production and describe how their daily chores now include cultivating the political fields.

Mark A. Grey provides an anthropological accounting of the costs and benefits of a meat processing plant on a small community in northwest Iowa. Exposing the shortcomings of traditional economic development measures that narrowly focus on job numbers and tax bases, Grey broadens the scope of economic development assessments by including social indices to provide a more accurate depiction. He details community social costs in terms of health care, education, law enforcement, and social upheaval created by the packing plant's employment practices. In so doing, he points out that the pork industry's drive to develop technologies that result in leaner, more standardized pork underlies the ability of packing plants to hire unskilled laborers at lower wages. These chapters demonstrate that the social consequences of industrialized agriculture that Walter Goldschmidt began documenting over fifty years ago continue largely unabated today. They also demonstrate the inadequacy of industrialization models that emphasize technology over people.

Chapter 1

Parma: A Story of Hog Hotels and Local Resistance

Laura B. DeLind

INTRODUCTION

I want to tell you a story. It is not my story, but one that has been told to me in various ways over the course of several years by those who lived it. It is the familiar, even ordinary story of rural development in which local environmental resources were used to generate short-term profit for external investors and to improve the state's GNP. It is also a story of NIMBY—Not in My Backyard—a brush fire of resistance by rural residents objecting to "development" in their backyards. It is the story of how the "old paradigm" of agriculture that John Ikerd describes in his chapter works against a rural community.

To tell the story in these ways mislabels the conflict and its resolution or accommodation and encourages us to focus on the catalyzing event rather than the underlying conditions that brought it about. It emphasizes pathways to consensus concerning the nature of the story rather than what it means to engage in local struggles. There is a tendency to tell these stories in a way that removes people and dehumanizes their experiences.

The story I want to tell will be less familiar because it puts people, context, and meaning foremost. It describes a specific conflict, but it also describes the contradictions and costs of local lives in conflict. The value of stories is that they alert us to previously unseen relationships and expand our scientific objectivity. Steven J. Gould notes that no story—and this includes our time-honored scientific ones—is without error. He cautions,

> But the solution to these errors does not lie in avoiding stories, for we do not have this option, given our essence [i.e., we are storytellers]. We must, instead, become more aware of the stories that underlie our methods and choices of topics for research, and we must learn to recognize the constraints and prejudices that any particular story must specify. We should, above all, enlarge our range of potential stories, for a choice among a thousand and one nights provides so much more scope than Cinderella (or another progressive tale of rags to riches) told every night (1995: 22–23)

Anthropologist Dianne Rocheleau makes a related point, "Perhaps the critical difference between 'mere anecdote' and significant story lies in the analytic content of the message and its synthesis of common content or process applicable to other circumstances. Both instructive stories and replicable experiments carry valid information for other times and places" (1991: 158).

These are my reasons for telling you the story of Parma and the hog hotels.

BACKGROUND

Project Feasibility

Jackson County is located in south-central Michigan about fifty miles from the Michigan-Ohio border. In mid-1983, a feasibility study for the Jackson County Hog Production Project (JCHP) in Michigan was completed (Allen Consultants, Inc. 1983). The $25,000 study, funded largely through the Michigan Department of Commerce, found that it would be feasible to build at least ten five hundred-sow hog production units in Jackson County, "giving an excellent return to investors and achieving the economic gains sought by Jackson County and the state" (ibid). These units, popularly referred to as "hog hotels," would be state-of-the-art, total confinement facilities producing some ninety thousand market weight animals per year. It was projected they would increase Michigan hog production by 8 percent, help eliminate the state's vast corn surpluses, and help Michigan's few remaining packers reach capacity with a steady supply of uniform, in-state animals for slaughter. These objectives were entirely consistent with the state's plans to reindustrialize agriculture (DeLind and Spielberg-Benitez 1990).

Equally feasible, from the study's point of view, was the way in which this $16.5 million project would be financed. Each unit would be organized as a subchapter "S" corporation. As calculated in the feasibility study, there would be thirty-two investors per unit and each investor would be required to invest approximately $51,600 per unit. As part of a subchapter "S" corporation, investors would be able to "deduct losses of the corporation from their personal income tax" (Allen Consultants, Inc. 1983; Willbanks 1985). A five-year depreciation allowance for single-use facilities would generate losses of over $11.3 million in as many years (Willbanks 1985). In addition, investors would hold $10 million worth of bonds issued by the Jackson County Economic Development Commission at 13 percent. The tax-free income from the bonds would contribute further to deductible corporate losses. Thus, investors would receive a double or compounded tax break. The bottom line, the study calculated, would be a net return of 24 to 27 percent to those with more than $50,000 to invest in it.

The Center for Rural Affairs has argued against such financing for environmental reasons and because it undermines smaller, family-scale hog producers who, while more efficient producers, cannot take advantage of the capital efficiencies enabled by the tax structure. Initiative 300, passed by public referendum in 1982, has made the corporate ownership of farmland and livestock illegal in Nebraska. There are no such restrictions in Michigan.

By contrast, the study gave little attention to the environmental impact of the ten proposed units. No direct mention was made of the over 134 million pounds of raw waste the facility would generate annually, the 150 million gallons of water it would require, suitable soil types or appropriate land-scapes (FEDF:na). One half-page of the forty-page document dealt with "Environmental Zoning and Siting." Here it was explained "that each unit would require a groundwater discharge permit from the Michigan Department of Natural Resources (DNR)" and that "units would be built only in agriculturally zoned areas with appropriate distances from any incompatible land use" (Allen Consultants, Inc. 1983). Waste handling would follow federal and state regulations and use the latest available technology to handle odor, which by all accounts would be "very minimal" (ibid 25).

If the study paid little attention to environmental and ecological issues, it paid almost none to the expressed needs and lives of area residents who would live with the facility. The one notable exception was a reference to several area farmers who had voluntarily offered to sell Sand real estate for future expansion. One of these farmers was a large landowner who had multiple business dealings with Sand. He sold Sand the original twenty acres for JCHP. A family-owned business had excavated the site. The facility's manure slurries were sprayed on his fields, and he rented his grain elevators to the Sand-owned Jackson County Grain Mill. Ultimately, he became an investor in the fifth confinement unit.

Several things need to be mentioned at this juncture. First, these "technical" and environmental assurances were provided by the same company—Sand Livestock Systems, Incorporated—that would be hired to construct and manage the facility. The study alluded to the fact that "if all went well" Sand would construct additional facilities throughout the state. Second, only three laws existed in Michigan that could regulate a farming enterprise accused of polluting the environment: the federal government's Water Pollution Control Act (PL92-500), the Michigan Water Resources Act (PA245 of 1929), and the Michigan Air Pollution Control Act (PA348 of 1965). None had been rigorously applied by the DNR, and prior to the Sand operation, only one national pollution discharge permit had ever been issued. Third, the DNR was seriously understaffed and its Water Resources and Air Pollution Control Commissions were unfamiliar with regulatory guidelines for animal agriculture, a condition due in part to "turf" struggles between the DNR and Michigan Department of Agriculture (MDA).

The Local Level

Parma Township is rural. In 1980 it had a population of 2,715 which dropped to 2,491 in 1990. There is no industry or major employer within the township. The village of Parma is a vacant "four corners" area with only three or four businesses still operating. Older residents tend to be farmers, laborers, tradespersons and service workers while newer residents are typically exurbanite or suburbanite professionals who commute to Jackson, Albion, and Kalamazoo for employment.

It was only with the "pouring of the cement" that most Parma Township residents became aware of their new neighbor, JCHP. Local concerns were raised almost immediately. What was being built? Why had the public not been consulted? How large a facility would it be? How would the pig waste be managed? What would it do to Rice Creek which ran only two hundred feet from the twenty-acre site? What would happen to the ground water, the air, property values? What environmental safeguards existed? How would this investor-owned facility affect smaller pig producers in the area? Should not an operation of this scale be considered industry, not agriculture, and regulated and zoned as such? These questions drew only assurances from representatives of Sand, the DNR, the Economic Development Commission (EDC) that there would be no problems.

An early petition signed by 290 area residents demanded a public informational hearing and attempted unsuccessfully to limit the number of animals that could be raised on farms within the township. Within two years, however, Parma and Springport townships passed zoning legislation that restricted intensive animal agriculture (*Springport Signal* 1986; Savitskie 1986a). While officially "on the books," a one-year, $1 million tax levy to enforce the ordinance was not approved by Parma Township residents (Carpio 1987).

It was not until early 1986, however, when it became apparent that JCHP intended to build a second set of five units and expand its operation into neighboring Springport Township that a "hog range war" erupted (Savitskie 1986b). By this time, area residents had firsthand experience with their new neighbor

JCHP and its three open air anaerobic manure lagoons (42 million gallon capacity) as a neighbor. Many reported a "horrific stench" that caused nausea, headaches, and respiratory ailments; burned eyes, noses, and throats; prevented sleep; and could be detected up to five miles away. Residents photographed dead pigs "piled up for days along the side of the road." Others reported pools of manure 8 inches deep lying on fields within 150 feet of private property. Airborne particles were thought to contaminate swimming pools. Manure spilled on roadways caused accidents. A JCHP employee became seriously ill from salmonellosis, a disease some felt he contracted "while working with dead animals at the Sand facility" (Willbanks 1986). There were also claims that neighboring wetlands and water resources were being compromised by manure runoff and field drains. Fish no longer swam in Rice Creek. One neighbor observed, "I've never seen such a drastic change in one year in that creek in my life. The creek now often has foam on it and is full of algae and alligator grass" (Weiker 1985).

Alarmed, residents redirected their complaints to Sand, the township planning commission, the county planning commission, the county health department, the DNR, the MDA, individual members of Congress, and the governor's office often on a daily basis. It was reported that within eighteen months "more than 400 complaints from virtually all 120 residents living within a mile of the hog hotel had been registered with the DNR" (Schmidt 1987b). Residents were told that there was no detectable pollution or violation of state building codes or waste discharge permits. From an official standpoint, the Sand operation was safe, legal, and unassailable.

Unsatisfied, Parma and Springport residents formed two grassroots organizations, an organization for lobbying and a nonprofit group for litigation. The lobbying organization, Save America's Farming Environment (SAFE) had some one hundred members, while the litigating organization, Farm Environment Defense Foundation (FEDF) was comprised of thirty to forty families. Despite these numbers, agricultural extension agents and agricultural scientists at Michigan State University continued to argue that "only a handful of residents made all the noise." While it is certainly true that some fifteen to twenty residents did the lion's share of the strategic planning and public speaking, there was a sizable though shifting support group behind them. The tendency to dismiss an issue on the basis of "small" numbers suggests that it is reasonable to quantify human rights and equally reasonable to overlook the economic and political conditions that marginalize segments of the population.

Together, members of these local organizations researched the tax codes, state and federal water discharge permitting regulations, specifications for lagoon construction, freeboard requirements, and agronomic rates for manure application and nutrient absorption. They developed their own maps, took their own photographs, collected their own data, and testified at state congressional hearings and before state review boards. They challenged the DNR's methods of water sampling and lagoon certification procedures. They argued the necessity of regular site inspections and test wells. Finally, they engaged legal counsel, hired expert testimony, and sued both Sand and the DNR (Water Resources Commission). The suit against the state was later dropped when the attorney general's office, acting on behalf of the DNR, joined the suit against Sand Livestock Systems, Incorporated.

For almost four years, these two organizations conducted a targeted campaign to keep the issue "alive" at the local level and in the state capitol, Lansing. I have told the story of this confrontation and the state and agro-food industry's ideological reframing of the issue around "right to farm" legislation elsewhere (DeLind 1995). These efforts eventually forced Sand to install a solid waste separator, recycle its waste water, and adopt strict odor management practices. An out-of-court judgment in 1989 placed further operational restrictions on the existing "hog hotels." In an allied suit, Sand purchased the home of an aggrieved neighbor. The litigation cost area residents well over $100,000 and put an end to Sand's expansion in Michigan. Another factor inhibiting effective participation in such matters is access to wealth.

In 1992 JCHP declared bankruptcy. Many feel that this bankruptcy was designed to coincide with scheduled depreciation and bond maturity. The official explanation, however, cited the cost of litigation, environmental constraints, and low hog prices (Cavins 1992). Today, thirty steel buildings and three lagoons stand empty on twenty denuded acres surrounded by chain link fence and mercury vapor lamps. Sand has gone on to construct and operate new hog confinement facilities in Wyoming, China, and Korea (PFRA 1993). The state, with the assistance of Michigan State University, is still advocating and addressing the needs of large-scale animal agriculture (DeLind 1991).

THE LOCAL IMMPACT

A Cost Accounting

This is only the briefest summary of one community's experience with corporately owned intensive animal agriculture. It suggests a David and Goliath battle in which ordinary people somewhat successfully confronted what they felt was a threat to their environment and quality of life. It seems to underscore the power of the people and the efficacy of democratic process. However, such a conclusion is seriously flawed. Little has been said about what the struggle cost the township or what it meant to those living there. Hindsight provides quite a different view of JCHP's impact and the degree to which the struggle and its resolution empowered the community to oversee its own development.

One way to asses local costs is to compare them against the initial development promises. Did JCHP purchase local corn and help reduce corn surplus in Michigan? Sand employees claimed the facility purchased anywhere from 3,200 to 5,000 bushels of corn annually but never disclosed how much was purchased from area residents. Area residents contend that JCHP trucked corn in from Iowa during the first several years of operation, and a production supervisor explained that Sand purchased all its high-value protein rations from Wayne Feed in Indiana, a subsidiary of Continental Grain. Furthermore, the Jackson County Feed Mill which supplied JCHP was also owned by Sand. Integration of this sort allows the income of one subsidiary business to become the expense of another. It stored its grain, not in an existing commercial elevator, but in elevators owned and leased to Sand by a large-scale farmer/investor. This pattern of input purchases appears consistent with the findings of a Minnesota study which compared such purchases between large livestock operations and small livestock farmers. According to the study, while livestock operations spent a certain base amount locally, regardless of size, the larger the operation, the greater the spending outside the local area (Chism 1993).

The original argument that JCHP, or any other large hog facility, would reduce corn surplus and improve the grain economy is suspect. Through existing federal commodity programs that encourage grain surplus and conventional, monocultural practices (Adams 1987; Strange 1988) the government, in essence, subsidizes cheap grain (or cheap raw materials) for value-added manufacture by the agro-food industry. Rather than reducing or eliminating price depressing surplus, operations such as hog hotels are its beneficiaries. Because they benefit from Michigan's surplus "problem," it is unrealistic to view them as a solution.

Stated another way, between 1978 and 1991 Michigan hog production increased by 71 percent (MSUAES 1992:7; PFRA 1993; USDC 1987). Yet, the industry still only utilizes 40 percent of the corn grown in-state, and Michigan corn prices in June 1994 were almost a dollar less per bushel than they were ten years ago (MSUAES 1992:12). Furthermore, while Michigan has seen an increase in the number of large-scale hog operations, hog slaughter has trended downward since 1984 (MSUAES 1992:8). The promised relationship between intensive hog production, an improved climate for grain farmers, and increased in-state processing has not materialized.

Did JCHP create local jobs? Yes, but few local residents filled them. According to Parma neighbors, Sand brought in its own construction crew from Nebraska. These "outsiders" stayed about two years installing Sand equipment, working with Sand machinery, and driving vehicles with Nebraska license plates. Once in operation, the facility employed twenty-two to twenty-seven people, a little over three persons per barn, according to a production supervisor. Of these, only ten to twelve were local residents (Kohler-Schepeler 1989; *Parma News* 1987). Equally significant, the higher paying positions of manager and unit supervisor were filled by persons from outside the Parma area. According to several Parma residents, Sand's local employees tended to be "rough" characters, unruly and physically aggressive. On a similar note, a production supervisor observed that local employees were unreliable and had a high turnover rate. It appears JCHP neither attracted quality workers nor provided quality employment opportunities locally.

It should be pointed out here that it is the elimination of labor, or more accurately labor costs, that constitutes a major economic advantage of large-scale, total confinement hog production. In fact, as Ikerd demonstrates in his chapter in this book and elsewhere, "The substitution of capital and mass-production technologies for labor and management is the primary advantage that large, specialized hog production units have over smaller, diversified operations. Large-scale, specialized hog production replaces people with capital intensive, mass-production technologies and centralized management" (1994: 3).

Locating JCHP in an area experiencing a 16.5 percent unemployment rate due to industrial layoffs was an opportunity for JCHP to exploit cheap labor, not for quality employment opportunities. Had the employment actually been the motivation, the state would have done better to provide incentives for smaller, family-scale hog producers. The fifty to one hundred farms, which would produce the hogs of a five thousand-sow confinement facility, would employ more people at higher paying and more dependable jobs.

Development rhetoric aside, it was not local capacity that was being built. Like Michigan's major hog slaughter facility, the vast majority of JCHP's investors were located in urban centers such as Detroit—as were its legal counsel, financial advisor, and underwriter/brokerage house. A complete listing of JCHP investors is privileged and protected information, unavailable to the public. Nevertheless, residents were aware that doctors, lawyers, judges, and industrialists had invested in the facility. Seventeen lawyers from a single Detroit law office with ties to the governor were reported to have invested nearly $1 million in the project (Schmidt 1987a). Profits created by the hog hotels were removed from the Parma area and along with investment capital were sheltered from taxation so there was no public benefit. Growth in state GNP due to increased hog production helped concentrate wealth in the hands of a few, not make it available to sustain the local community.

Less obvious, but no less real, were the opportunities Parma lost as a result of JCHP. A family-run antique apple orchard and historical museum, for example, employed four local residents full-time and hired a seasonal crew of as many as twenty-eight locals. Far from being intrusive, the enterprise preserved and celebrated local culture, history, and the natural environment. It utilized sustainable methods of apple production and relied on a spring-fed lake for irrigation. They were planning to plant an additional 40 acres and hire a wine maker experienced in brewing French apple cider, when odors from the hog hotel, heated confrontations with JCHP management, and severe environmental damage caused the family to abandon the business and permanently leave the area. The excavation of a "farm drain" (3.5 miles long, 20 feet wide, and 12 feet deep) in anticipation of JCHP's expansion diverted the spring and caused the orchard's irrigation lake to become unusable. A lawsuit initiated over this environmental alteration was ultimately settled in 1991 in the family's favor.

Property values, particularly those around the JCHP facility, also collapsed. As one realtor confessed, "many people are very hesitant to purchase a home in that area because of the possibility of the Pig Hotel and the uncertainty of the possibility of more of them" (Colgan 1984). A recent study appears to corroborate this relationship, reporting that hog operations can depress the sales value of neighboring homes and real estate (Palmquist, Roka, and Vulkina 1995). The $100,000 that Sand Livestock Systems, Incorporated, claimed it paid annually in local taxes looks considerably less impressive when compared to the costs assumed by area residents. In many ways, the local economy was held hostage to outside interests.

A Human Account

However important, these are *not* the local costs or losses that I most want to talk about. A revised econometrically based, cost-benefit analysis patterned after the original feasibility study does little to recast our basic thinking about the nature of the Parma experience. It insists that reality be quantified and have statistical significance. We fall into a reductionist logic that permits us to reconcile an improved market share in pork with lost property value in Parma, or ten displaced families in Jackson with eighty retained jobs in Detroit.

Underlying such an orientation is the unspoken assumption that one "voice" carries authority and provides the most appropriate basis for socioeconomic assessment. This authoritative voice, furthermore, remains external to the reality it describes. It is superimposed upon place, and it discounts human experience, emotion, and meaning in favor of tidy, scientifically manipulated, cause and effect relationships. It is used repeatedly to justify the interests of power, financial interests, and the state. It also tends, in the name of rationality and neutrality, to deny the validity of other voices and their struggle, something that figures at the heart of all grassroots efforts.

I examine the very thing such assessments do not see—the human dimension. The discussion that follows is based heavily on in-depth interviews with twelve area residents, all of whom actively fought JCHP and the expansion of hog hotels in their own backyards. It is meant to give some expression to the human experience that underlies cost calculations. No economic cost-benefit assessment, for example, could ever account for the fact that three of the twelve residents cried when recounting their stories, or for the equally poignant admission, "It's hard to recall five years of hell." These most human of realities are stripped away as externalities by economic assessments.

For many area residents, and for these twelve people in particular, the existence and operation of JCHP was a sustained insult to their environment and their integrity. They felt deliberately bypassed in the original decision-making process and deliberately deceived. "He [Sand] told us there would be no smell, that we would never know the operation was there. He lied through his teeth. He'd already been run out of Nebraska. And I believed him. I had no reason not to, until the first day I smelled the stench." One DNR employee is similarly remembered and deeply resented as "a person (who) could smile at you and lie to you at the same time. He just out and out lied to us."

The anger and frustration of not being dealt with honestly was strengthened by what residents interpreted as public indifference to their problems and a disrespect for them as individuals. Reflecting the sentiments of many, one activist asserted, "Sand was pushed down our throat, as far as I'm concerned." Another recalled, "I spent the entire summer [1985] on the phone trying to find an attorney who would be willing to talk to us. We found no one, and moneywise we couldn't afford it." A state representative who visited the area and vowed assistance, later wrote a letter explaining that he "could not find any information prohibiting this type of operation because of environmental concerns" (Griffin 1986). This signaled the end of his involvement, though he tactfully added, "If any other information comes my way, I will forward it to you." Apparently, nothing ever did. Likewise, DNR repeatedly assigned inexperienced field staff to the case who knew little about intensive animal agriculture. "We had to teach them how to do their jobs."

The evidence residents submitted for governmental review was devalued and often disregarded. Their direct observations of the deterioration of Rice Creek were inadmissible because there was no official documentation of its prior condition. Phone calls to the DNR complaining of pig manure floating in Rice Creek encountered the demeaning response, "How do you know that it's *pig* manure? Describe it to me." The discrepancies residents found by recalculating groundwater table elevation, lagoon storage capacity, and the acreage actually available for manure application were dismissed on the advice of, or in deference to, "experts," several of whom were employed by Sand or had ties to other intensive animal operations (Willbanks 1986). This dismissal was doubly galling because a number of local residents were themselves professional engineers and technicians. By contrast, official determinations often went unsubstantiated. Resident recall, for example, that an eleven-inch rise in the level of Rice Creek in a single, rainless day was attributed by the DNR to "roof runoff." Official calculations often were rubber stamped, passing with little critical review from one office to the next. The calculation for water table level, for example, as made by the Sand engineer was accepted site unseen by the DNR. This despite the fact that the engineer admitted not knowing where the Rice Creek bench mark was located.

Prior to the "hog hotel" controversy, most Parma residents shared a sense that government was there to serve their needs. They were respectful of the law and the processes that created and upheld it. Though few had any political experience or "know how," most felt that they only had to express themselves to get what they wanted. With the introduction of the "hog hotel," they began to sense that the system was reluctant to embrace their interests over those of power and profit. They began to feel that collusion, if not outright conspiracy, existed among politicians, government employees, and Sand supporters. "Many high up political persons were invested in the operation. No one wanted to look too closely or to step on toes. It was a case of 'one hand washing the other.' For a long time we were laboring under the illusion that if we got to the right person, he would say: 'Wait a minute Sand! You can't do this to the people of Parma.' That person does not exist" (Flory 1989).

As this realization grew, so did local anger over the vulnerability of area residents and their loss of real security. Residents realized that they would have to protect themselves and they had no clout to do so as individuals. "We formed a corporation to fight a corporation. You can talk all you want, you can present all the evidence you want, but nothing will happen, no one will tell the truth until you go to court." The prospect of going to court was daunting and expensive, but for many residents it was absolutely necessary. The fight was over something as essential as their homes—the spaces and places that gave identity and meaning to their lives. This was their bottom line. "I loved it there. [Yet] I couldn't walk in the woods. When you can't breathe the air in your own home, when you're sick, what can you do? If you can't live in your own home, then what's important? I was not giving up my home. I was just so angry. They were taking everything I had worked for away for their profit."

The very existence and identity of the most active residents became bound up in the fight. The conflict became totally black and white. "You were either for us or you were against us." This polarization of issues was devastating for everyone. As one woman recalled,

> It consumed me. It was all I could think about. All I could talk about. I was interested in nothing else, but shutting them [Sand] down. I was totally obsessed, but I feel I had every right to be. Inhumane invisible intruders lived with us all those years. It created drastic injustice to anything in its path, to the trees, to the land, to the water, to the air, to the rabbits, to the birds, to the animals, to the people in Parma. . . . No one should have to live under these circumstances. I hated everyone who wouldn't help in one way or another. I was angry all the time. It's hard to be angry all the time. It destroyed me.

The tension created within the community was palpable. Meetings, phone calls, research, and testifying consumed upwards of forty hours a week, week after week, month after month. It became a full-time job and competed with the needs of family, work, and friends. Children were neglected. New businesses were put on hold. There was no time for socializing. Husbands and wives argued over the totality of their respective commitments. People began burning their candles at both ends, staying up late, eating on the run, and running all the time. Tempers flared. Neighbors and long-time residents caught in the crossfire no longer spoke to one another.

As residents fought the "hog hotel," Sand employees and local Sand supporters fought back, harassing and intimidating residents in turn. Cars were followed home from meetings, and people were run off the road or followed as they walked near the facility. Mysterious phone calls were made in the middle of the night as were the taunts, "You'll never get rid of us." Trespass was frequent. Windows were broken, chickens killed, a new farm fence was cut, and a night watchman was assaulted. Residents responded to this last incident by forming a local vigilante group. Armed with guns and baseball bats and sustained by adrenaline, they spent the next night searching for intruders, though thankfully they found none. A small rural community was transformed into a state of paranoia. People began to fear for their own and their families' physical safety. They believed their homes were under surveillance and their telephones tapped. They suspected the "enemy" was not beyond committing murder, and they too admitted giving the possibility some thought.

Residents were well aware that they were being consumed by the conflict, yet they were unable to disengage from it. It had acquired a momentum of its own. It became a matter of principle, of personal integrity. They publicly promised to stand up to the arrogance of money and the privilege of power. Dozens of area residents contributed between twenty dollars and one thousand dollars each toward a legal defense. Residents felt they had a responsibility to the community and especially to those least able to protect themselves, such as children and the elderly. They could not back down.

However, the intensity and duration of the commitment resulted in tragedy and inestimable personal loss. The most obvious was the stress-induced death of the movement's thirty eight-year-old leader who died of a heart attack, leaving a wife and two young children.[1] Grief and anger further galvanized the struggle. "You don't forget about this stuff. The guy's buried across the street there. I mean every day I come out here and see his grave and think about it. Every day, you know. . . . He shouldn't have had to be involved doing what he was doing."

CONCLUSION: WHO LOSES?

The judgment against Sand came less than two years later. Then, like now, no one expressed satisfaction or joy over the decision. Many of those who were most active had already left the community. Others only wanted to get on with their lives and "put it behind them." They admitted to being physically and psychologically exhausted, even "broken," by the struggle. "I've become an old man," reflected one resident. "It destroyed my life," confessed another.

For residents, the losses far outweighed any positive outcome. To a person, they admit that the grassroots effort was not worth the sacrifice. "We did stop Sand from building additional units, and I feel good about that. I believe we were the greatest opposition they had ever had." Nevertheless, "There was more lost than was ever gained—personally, financially, and environmentally. [Things will] never be the same. They robbed Parma big time." Underlying these sentiments are other equally enduring losses.

Local residents expressed a loss of faith in the state. "Our own state government really let us down. People who think they can fall back on the state for protection need to wake up" (Flory 1989). At the same time, a few individuals felt they would take part in a community-based struggle again. Though somewhat embarrassed, most admitted that they would move rather than take on city hall or fight big money again. Others said they no longer have the energy or are willing to make the personal sacrifice. Yet another conceded, "There really is no point. Money talks, and money gets what it wants, regardless."

And that is how the story of "Parma and the hog hotels" ends. The ending is hardly a happy one. As a conclusion it is far less promising than it might have been in terms of human capacity building and sustainable social and environmental development. In this regard, Parma's losses are our own. We are all diminished by the ultimate weakening of people at the local level. It is sadly ironic that in the process of resisting external and exploitative forces, Parma residents were estranged from their past and blocked from assuming active roles for their future. Parma is a story of ordinary people engaged in a heroic struggle. But far too often the struggle is itself defeating.

NOTE

1. It appears that the death of a local activist due to the stress of fighting corporate agriculture is not a unique occurrence. A similar and equally tragic death was reported in North Carolina (Cecelski and Kerr 1992).

REFERENCES

Adams, Jane. 1987. Business farming and farm policy in the 1980s: Further reflections on the farm crisis. *Culture and Agriculture* 32: 1–6.

Allen Consultants, Inc. 1983. *Jackson County hog production project: Final report.* Lansing, Mich.

Carpio, Ninia L., "Parma Voters Reject Levy for Hog-hotel Fight," *Jackson Citizen Patriot,* 13 May 1987, A-3.

Cavins, Denise B., "Two Hog Farms May be Folding," *Jackson Citizen Patriot,* 25 January 1992 A-1, A-2.

Cecelski, David, and Mary Lee Kerr. 1992. Hog wild. *Southern Exposure* Fall: 8–15.

Chism, John Wade. 1993. Local spending patterns for farm business in southwest Minnesota. Masters Thesis, University of Minnesota Department of Applied Economics, St. Paul, Minn.

Colgan, Craig, "Parma Planners Reject Hog Petition," *Jackson Citizen Patriot,* 19 August 1984, p. na.

DeLind, Laura B. 1991. Sustainable agriculture in Michigan: Some missing dimensions. *Agriculture and Human Values* 8(4): 38–45.

———. 1995. The state, hog hotels and the "right to farm": A curious relationship. *Agriculture and Human Values* 12(3): 34–44.

DeLind, Laura B., and Joseph Spielberg-Benitez. 1990. The reindustrialization of Michigan agriculture: An examination of state agricultural policies. *The Rural Sociologist* Summer: 29–41.

Flansburg, James, "The Sweet Breath of Spring" (The Old Reporter Column), *Des Moines Register,* 9 May 1995, 11A.

Flory, Bradley. "Parma Township Seeks 1-Mill Levy for Legal Fees," *Jackson Citizen Patriot,* 13 January 1987, p. na. "Them fumes' Revolted Parma Neighbors," *Jackson Citizen Patriot,* 25 June 1989, A-1, A-2.

Farm Environment Defense Foundation (FEDF) Fact Sheet.

Gould, Stephen J. 1995. Speaking of snails and scales. *Natural History* 104(5): 14, 16, 18, 20–23.

Griffin, Michael J. Letter to Mrs. Gilbert Cockroft. 24 February 1986. Lansing, Mich.

Ikerd, John E. 1994. The economic impacts of increased contract swine production in Missouri: Another viewpoint. Columbia: University of Missouri.

Kohler-Schepeler, Jeanine, "Neighborhood Corn Bank Thriving," *The Parma News,* 10 August 1989, 1, 3.

McLemee, Scott. 1995. Public enemy. *In These Times* 15 May: 14–19.

Michigan Farmers Union (MFU). Personal communication with Carl McIlvain, president MFU, 20 June 1994.

Michigan Pork Producers Association (MPPA). 1995. *Michigan Pork Producers Association News.* Feb/March.

Michigan State University Agricultural Experiment Station (MSUAES). 1992. *Status and potential of Michigan agriculture—Phase II: The swine industry.* Special Report #45, October. East Lansing, Mich.

Michigan State University News Bulletin, "Center Seeks to Improve State's Animal Industry," 3 June 1993, 6.

Palmquist, Raymond B., Fritz M. Roka, and Tomislav Vulkina. 1995. The effect of environmental impacts from swine operations on surrounding residential property value. Paper. North Carolina State University Department of Economics, Raleigh, N.C.

Parma News. Comment in the "News Box" (letter from the farm manager for Sand Livestock Systems, Inc.), 7 May 1987, p. na.

PrairieFire Rural Action (PFRA). 1993. *Hog tied: A primer on concentration and integration in the U.S. hog industry.* Des Moines, Iowa: PFRA.

Rocheleau, Diane E. 1991. Gender, ecology, and the science of survival: Stories and lessons from Kenya. *Agriculture and Human Values* 8(1&2): 156–165.

Savitskie, Jeffrey, "'Industry' Regulated: Parma Pig Ordinance Will Control Odor, Waste, Even the Squeal," *Jackson Citizen Patriot,* 8 May 1986a, p. na.

———. "Hog 'Range War' Heats Up in Parma," *Jackson Citizen Patriot,* 29 March 1986b, A-1, A-2.

Schmidt, Wayne A., "Blanchard Eyes Plans for 20 Hog Hotels," *Jackson Citizen Patriot,* 16 December 1987a, A-1, A-2.

———. "State Sues Hog Facility Over Odors," *Jackson Citizen Patriot,* 10 September 1987b, A-1, A-2.

———. "Weak State Environmental Law Triggers Violations," *Jackson Citizen Patriot,* 24 February 1988, A-1, A-2.

Springport Signal, "1986 Public Notice: Springport Township Zoning Ordinance," 8 September 1986, p. na.

Strange, Marty. 1988. *Family farming: A new economic vision.* Lincoln: University of Nebraska Press.

Thumb Farm News (TFN), "Lower Thumb Grain Markets," 1 June 1994, 7.

Weiker, James, "County Aiming to Hog the Market," *Jackson Citizen Patriot,* 13 October 1985, p. B-1.

Willbanks, Michael L. 1985. Presentation to the Michigan House Agricultural Committee. 3 April, Lansing, Mich.

———. Letter to Jeffery (sic) K. Haynes, 2 July 1986. Albion, Mich.

U.S. Department of Commerce (USDC). *1980 Census of Population: General Population Characteristics—Michigan (Part 24)*. Bureau of the Census, Washington, D.C., 1980.

———. *Census of Agriculture—Michigan (Part 22)*. Bureau of the Census, Washington, D.C., 1987.

———. *1990 Census of Population: General Population Characteristics—Michigan (Part 24)*. Bureau of the Census, Washington, D.C., 1990.

Chapter 2

Inside the Industry from a Family Hog Farmer

Jim Braun with Pamela Braun

Life is changing drastically for me and my family as we seek to find a way to hold on to our family farm in order to pass it on to our children. At times it seems that the American farm dream may no longer be attainable by the time our children are grown. I have been an independent farmer in north-central Iowa for nearly twenty years now. In 1990 my wife, Pam, and I bought the family farrow-to-finish hog operation that currently markets ten thousand hogs each year. For almost twenty years we and my parents have grown corn together on eight hundred acres of some of the finest soil in the world. For sixteen of those years Pam and I have raised our three children and participated in their local school, church, and community projects. Ball games and recitals, harvests, and hogs have been our way of life.

It is from here, in north-central Iowa, that I watch what was once a thriving source of income for the multitudes of independent farmers become profitable for only a privileged few. From here I watch an industry in turmoil over its policies and place within our communities and our state. From here I watch the dominant farm organizations, publications, and land grant university groups tout the excellence and superiority of the newly emerging mega-hog farm elite while hotly dismissing any claims to the contrary. I wonder how these mega-hog conglomerates arrive at their "low cost and high-efficiency" figures cited around our state universities, legislative floors, and local co-ops. What kinds of subsidies like those Laura DeLind describes in Michigan are offered to entice the pork conglomerates into Iowa, to enable them to remain financially viable, and to defend their social and environmental indiscretions? What is their impact on the lives of our rural farm families and our communities, as well as each

Iowa taxpayer? Where does my family go from here? These are the current questions my wife and I ponder each day of our lives inside Iowa's changing hog industry.

FROM GENERATION TO GENERATION

My great-grandfather immigrated from Germany in the mid-1800s, settled in Illinois, and then purchased a farm in north-central Iowa. In 1920 his son, my grandfather, began working the family farm and raising hogs. For years he and my grandmother raised hogs, crops, gardens, and two children in a manner similar to what Blaine Nickles describes in his chapter. After my father and mother were married in 1948, they joined my grandparents in farming. My first memories of hogs date back to about 1956, when at the age of five I sat on an overturned five-gallon bucket watching sows farrow their litters. Breeding stock and the arts of crop production and animal husbandry were passed from generation to generation, father to son.

From 1959 to 1969 my family raised hogs each summer by farrowing sixty sows in individual huts out in our pasture. Dad would then buy six hundred feeder pigs to finish out on the concrete yards and in the barns by the house. The yard pigs were ready for market and sold by fall. This allowed room for the field-farrowed pigs to be moved into the barns by the time cold weather arrived. The added protection of the barns was necessary for the young hogs while they continued to mature for market. Sows were not farrowed in the winter because the old buildings were not warm enough for young piglets to survive.

This farrowing and feeder-pig cycle continued for a decade until the late 1960s brought a disease called MMA (Mastitis Metritis Agalactia), which caused sows to lose their milk and their young starved to death. The only treatment was a series of shots strategically timed immediately after farrowing. If the sequence was missed, the piglets died. The summer pasture farrowing method made giving sows their scheduled shots extremely difficult. Even the tamest sows became very leery after receiving the first shot, and thousands of field-farrowed piglets died.

In order to solve this and other problems in hog production, my father began making a major change in our hog operation in 1970. A concrete pit was built, and concrete slats were installed to service a 144 foot by 44 foot farrowing house that was totally enclosed. Furnaces were installed for heat in the winter, and exhaust fans were installed for ventilation in the summer. An automated feed system was installed to bring feed to each of the 104 individual sow stalls set in four rows. Each stall was its own self-contained sow hotel, with an automatic feeder, waterer, and manure removal system. We farrowed year round, and the sows could not run from their shots, thereby helping to ensure the health and safety of the piglets. By the fall of 1974, six more buildings were added, and all of my father's hogs were on slatted floors and under aluminum roofs. Hog confinement technology was in its infancy, and we were on the cutting edge.

We had to learn many things the hard way. Issues of comfort and safety for hogs, disease control, and manure management were all new and complex in this type of production. We constantly applied creativity and innovation, and for each new problem we found a solution. Confinement solved many problems associated with hog production. The pigs were protected from the elements, which increased their feed efficiency and their rate of gain. Sow productivity was increased because they could be weaned and rebred to farrow no matter the season or weather. Also, left on their own outside, hogs develop a social structure and a pecking order that is rigidly enforced. Only those at the top of the hierarchy thrive. They receive the larger portions of feed by bullying the smaller and weaker hogs. Stronger and more dominant pigs mutilate and often kill weaker and smaller pigs. Grouping hogs into smaller, protected numbers inside helped to reduce the "Boss Hog" syndrome. In many ways, confinement was advantageous for the hogs as well as the farmers.

However, confinement units have their disadvantages as well. The significant amount of capital needed for the initial construction of confinement buildings required increased efficiency from the herd to meet principal and interest payments. The dust and gases in the building that Kelley Donham describes in his chapter deteriorated the equipment, and maintenance became a continual cost of production. Utility costs became monthly expenses. Vaccination and antibiotic costs rose because the increased concentration of hogs created an environment conducive to the rapid spread of diseases, similar to the condition often faced among groups of young children at day care facilities. Because of the intense labor involved in this type of production and the pressures of continual farrowing, hired employees became a necessary part of the operation. These overhead factors greatly increased the cost of confinement production. The larger the confinement unit, the greater the fixed overhead costs, the greater the problems with disease, the greater the problem of proper manure management, and the greater the problem of managing the pig flow. Efficiency seemed to be essential to maintain profitability.

Another important issue is that since hogs are very social animals, the people who own and manage hog confinement units must find employees willing to commit their lives to working peaceably with very stubborn animals whose productivity depends in part on the attitudes of those caring for them. Finding employees who can work well with hogs, as well as with other people within the unit, is a challenge. Twenty-seven years of experience in confinement has taught me that raising hogs by this method is not easy or problem-free. The people in hog management have always been important, but in confinement pork production they become essential.

THE TURNING POINT

During the spring of 1994, a local high school student asked me to participate in an interview for public television. The student wanted my reaction to the growing number of large corporate hog confinement facilities being built in our area. He wanted to know what I felt their impact would be on the independent hog farmers in the state. I said that I was not concerned with the large hog units springing up and surrounding our county like creeping charlie. I said that when the price of pork fell from overproduction, these mega-operations with their high overhead costs and low efficiency rates would go broke. At that point, independent producers would buy the mega-producers' buildings for twenty cents on the dollar and raise hogs in them at a profit. However, many things happened that summer that caused me to wonder whether or not what I had said was true and whether the free market system which I had been taught about in grade school was indeed governing the hog industry. I now see that I was naive.

The first indication of serious problems came from the major farm publications later in the spring of 1994. Two issues kept cropping up and were being declared as the most essential factors determining who America's future pork producers would be. These factors were "access to capital" and "relationship." In the past, access to capital meant going to the bank, telling the banker your plans for hog production, and getting money for the loan. Since hogs have always been the mortgage lifters for farmers, pork production was a good business venture for the farmer, the banker, and the community. Getting "access to local capital" was never difficult. "Relationship" always referred to whom your ancestors were and your reputation in the community. It was not until the spring of 1995 that I began understanding the true meaning of these factors and their implications for independent pork producers. This understanding came one step at a time.

The next step toward understanding industry changes came after a visit with an old friend who was very influential in federal agricultural policy. I asked him what he was seeing in national pork production trends, especially the expansion of large units across the nation. He said, "I don't know, Jim. It's smart money being invested." Through the course of our conversation I realized that efficiency and cost of production were not going to determine who tomorrow's pork producers were going to be. He was told by a large integrator that it would not be a question of who could raise hogs at forty dollars per hundred weight and survive, but who could raise hogs at thirty dollars per hundred weight and survive. I said, "You know that no one can raise hogs at that price and survive." He replied, "I know that, and you know that, but that's what this large integrator said." I realized that this was not the free market system as I had been taught or hog production as I knew it.

During the fall of 1994, hog prices fell to a twenty-two-year low. Farmers received comparatively less for their hogs during the fall and winter of 1994 than during the Great Depression. Independent farmers were liquidating their herds in droves. A representative from the Iowa State University Extension Service told me that 45 percent of Iowa's independent sow herd was taken to market from the fall of 1994 through the spring of 1995. The USDA *Hogs and Pig Report* reported a 22 percent decrease in Iowa's sow herd. During that same period there was a 23 percent increase in large-scale industrial sow numbers in Iowa. While independent farmers were liquidating, the concrete trucks never stopped pouring concrete for the corporate sow. While the farm publications and land grant universities were telling farmers how to salvage as many assets as possible through their liquidation process, the concrete trucks never stopped pouring for the corporate sow. As my wife and I tried to calculate how long we should endure low prices before liquidating our herd, the concrete trucks never stopped pouring concrete for the corporate sow. I wondered what the corporate hog producers knew that I did not.

At this point, I began to hear rumors of differences in pricing structures. A report in *Hogs Today* magazine reported a difference of twelve dollars per hundred weight paid to the large producers in North Carolina over independent producers. This looked like an economic inside "fix" to go along with the inside political "fix" in North Carolina that Senator Robert Morgan discusses in his chapter. One of the large corporate producers in Iowa had a contract guaranteeing him a base price of forty-six dollars per hundred weight with an open top plus premiums. This meant that he would never receive less than fifty dollars per hundred weight and if the market went up he would receive a raise equal to the market price. But if the market went down even as low as twenty-six dollars per hundred weight, as it did in the fall of 1994, he would receive no less than fifty dollars per hundred weight. This was an economic advantage to which the independent farmers did not have access. A guarantee such as this would have kept any independent producer in business. Agreements such as this would justify the corporate producers' optimism and desire to keep building even though hog prices were forecast to average thirty-six dollars per hundred weight for all of 1995.

I could now understand why the corporate producers were continuing their expansion. What I could not understand was why the packing plants were willing to pay a select few producers above-market prices for their hogs. Then I received an interesting article from a cattle magazine in May 1995, which I believe answered my questions. Packing plants were making purchases both from producers under contract and from noncontracted producers. Producers under contract provide approximately one-third of the animals packers killed per day. The packers say that these contracted animals provide the highest quality meat, therefore they pay these producers the highest prices. Federal price reporting laws state that if a packing plant reports to U.S. Market News Service one price paid to any producer, they must report the high and the low price paid to all producers on that same day. The U.S. Market News Service then releases the average of those prices to all news broadcasting stations. This ensures that all producers know the average price that was paid the previous day for the livestock sold. By this system producers can determine whether they are getting a fair price for their livestock. However, by not reporting the contracted one-third of their purchases, the packing plants can greatly lower the price that they pay for the other two-thirds of their noncontracted livestock purchases. Also, contracted livestock gives the packing plants a captive supply of slaughter animals, which decreases their need to bid competitively on the open market. This allows them to further control the prices they pay to producers. These simple livestock purchasing maneuvers allow the packing industry to subsidize corporate hog expansion and at the same time increase their own profits. All of this is being done at the expense of noncontracted independent hog farmers.

All packing plants in the Midwest reported record profits following the price plummet in the fall of 1994 and the spring of 1995. The corporate concrete trucks never stopped running, and independent producers kept running too, away from hog production.

WHAT WAS A FARMER TO DO?

For generations, tens of thousands of us farmers relied on pork production to put food on our tables, pay for our land, and help pass our land on to our children. For generations, we pork producers went to town to worship, to educate our children, to buy supplies, and to entertain ourselves. Rural communities thrived as farmers thrived. I worked to provide the same opportunities for my children that my parents and grandparents worked to provide for me.

My grandfather, George Washington Braun, Jr., always said, "As long as you stay at home and take care of your business, the bad times will come and the bad times will go, but your business will remain." He practiced this diligently during the Depression and throughout his farming career, and it served him very well. My father committed all of his energies to the farm and little else, and farming went well for him. I planned on pouring myself into doing the best possible job of raising crops and hogs and planned on the same success based on my own knowledge, expertise, and hard work. But everything I was learning about the changes in the hog industry in Iowa and the nation led me to believe that good management, excellent genetics, and the use of current methods and technology alone would not lead to profitability for independent farmers such as myself. It was time to take action outside the confines of the combine, hog buildings, and computer print-outs.

I was concerned about the injustices I was discovering in the hog industry and their long-term effect on my farm and our rural communities. In the fall of 1994 my wife, Pam, and I frequently discussed our future as hog producers. Even with a debt-load that we had worked very hard to reduce to a minimum, losing substantial money on each hog we sold at the packing plants was rapidly eroding our equity. We asked ourselves many hard questions. How soon should we consider liquidating our herds before our equity was entirely gone? How much of the eroding equity should we allow for taxes to pay "capital gains" resulting from the sell-out? Where would we turn next? We watched our farm friends all around sell their sow herds and drop out of production altogether or give in to the false promises of contract feeding. Our rural communities groaned under the weight of unfair prices in the marketplace while industrial expansion burgeoned. Something needed to change quickly.

I began looking off the farm for those who could help return justice and common sense to the industry. My first step was to call the Iowa Pork Producers Association (IPPA) and the National Pork Producers Council (NPPC). These organizations were founded by cooperative efforts of family farmers who raised hogs. The intent of these organizations was to promote profitability and excellence in the pork that family farmers were raising. In 1986, pork producers throughout the nation voted to adopt a mandatory federal check-off fund from each hog sold to finance pork promotion efforts. The IPPA and NPPC were my representatives to the world in order to ensure fair pricing at the packing plants, to promote our products to consumers, and to advance the quest for accurate research to help family farmers do our jobs better. At every turn, in every conversation, I grew to believe that the IPPA and the NPPC were no longer my representatives but were now representing the large corporate-owned producers. I was continually told to adopt new technologies, know my cost of production, and provide better management. I had done all of these things for twenty years.

Later in 1995, Marty Strange from the Center for Rural Affairs—a rural advocacy organization in Nebraska—fellow producer Kelly Biensen, and I discussed growing pork producer concerns with Larry Graham, then C.E.O. of the NPPC. When I expressed my concerns about discriminatory pricing structures at the packing plants, Larry replied in astonishment that the NPPC had worked very hard to establish "a good working relationship with packers," and they did not want to discuss any matters that might hinder that relationship.

I came to realize that the only way to awaken the IPPA and NPPC to the growing producer discontent was to become active in their leadership. My check-off money was exacted to educate farmers and consumers, encourage research, and enhance promotion of family farm production of pork. Heightened involvement within the IPPA provided me greater opportunity to recover those funds for their original purpose. In the winter of 1997 it was discovered that NPPC had gone so far as to hire a Washington, D.C., consulting firm to conduct background checks on organizations promoting family farming to assess whether they could be "influenced." I intended to hold the IPPA and the NPPC accountable for misuses of their authority and what I viewed as illegal uses of our money.

My efforts over the next few years met with constant resistance. I believe this took place because these organizations were working together with the packing plants and large integrators to consolidate the industry into fewer hands. I arrived at this conclusion from an incident widely reported in July 1995. A memorandum from Larry Graham, head of the NPPC, was sent to twenty-eight of the nations' largest pork producers, including Murphy Farms, Premium Standard Farms, Tyson Foods, Cargill, Continental Grain, Iowa Select Farms, Heartland Pork, and other large-scale producers, inviting them to a meeting in Washington, D.C. The goal of this meeting was to acquaint these companies with the organization's legislative agenda and to familiarize them with their newly employed "heavyweight Washington, D.C.-based media consulting firm to work with [us] in trying to change [this] perception" that "living next to a hog farm is bad" (Graham 1995). He also stated the NPPC's intention to acquaint these twenty-eight businesses "with a major new NPPC effort to influence policy makers in Washington, D.C., regarding environmental issues." Not one representative of small, independent family farmers was invited to this meeting. I was mistaken in my earlier assumption that the IPPA and the NPPC existed for the benefit of my business.

After my early attempts to find support from the IPPA and the NPPC failed, I turned to my U.S. congressmen and my state legislators. My congressmen merely told me that these were changing times. My legislators told me that their hands were tied by the "ignorance of urban legislators who refused to listen." I was repeatedly told, "Don't stand in the way or the steamroller will run right over you." It seemed that these bastions of democracy were telling me that there was no hope or recourse left within the democratic process. The deeper I probed, the more I began to see the tie between the politicians and their campaign funds that are so generously offered by industrial agribusinesses. For example, the owner of Iowa Select Farms, one of the largest industrial hog producers in the country, contributed $48,000 to republican Governor Terry Branstad's 1994 election campaign. It was later discovered in the winter of 1995 that a number of republican legislators had received additional funds from Iowa Select through an "employee fund" set aside from employee salaries. When this news broke in the newspapers there was a clamor among newly elected officials to return those campaign contributions and dismiss any obligatory connections with Iowa Select.

In the winter of 1995 I contacted Marty Strange at the Center for Rural Affairs in Walthill, Nebraska. People at the center clearly understood the issues I was discovering and were eager to help. That help took the form of a meeting that January among a large group of concerned farmers and other rural citizens from across the state and Marty Strange. The majority of attendees were hog farmers actively involved in a variety of organizations trying to guide the industry's direction. Our goal was to assess the forces driving industry transitions and then to discern what steps should be taken to effect constructive change. Out of this meeting emerged Friends of Rural America (FRA), and I became its vice-president.

With the support of the Center for Rural Affairs, we began publishing a newsletter, *The Voice*. Our goals were to promote real prosperity in our rural communities; true integrity for our land, water, and air; and lasting social and commercial ties which support our towns. We hired farmer-member Mike Sexton to be our eyes, ears, and influence at the statehouse. We felt the present course of the Iowa legislature was misguided. It was our belief that they were receiving their direction from the governor's office, the Iowa Farm Bureau Federation, the Iowa Pork Producers Association, Iowa State University, the packing plant lobbyists, and the pro-mega pork lobbyists which abounded at the statehouse. All of these groups seemed to have more influence than family farmers. As I began to ponder the question of why this would be true, I was again reminded of the fact that individual family farmers do not have PAC money to contribute to campaigns.

No one I know can understand why the Farm Bureau is taking a stand in favor of industrialized pork production. Many county Farm Bureau associations have passed resolutions to pass on to the state Farm Bureau level that would bring some sanity to the issue, but when they get to the state level of the Farm Bureau they are somehow turned into nonissues. The motivation of Iowa State University, our agricultural land grant institution, is much easier to understand. University funding is under the potential budget knife from the Iowa legislature each year. This causes the university to seek outside funding sources. Agribusiness seems to be more than willing to take up the financial slack. I wish I were altruistic enough to believe that agribusiness believes in unbiased research and wants nothing in return for the funds it is investing. The pressure exerted to silence faculty who do not support the industrial hog agenda indicates that this is not the time for such altruism.

Groups of concerned citizens began growing in every community where large-scale hog facilities emerged. These groups all seemed to try the same channels we had, including all of their local governmental agencies. What they found was a lack of any rational recourse to have their concerns and needs met.

I was asked to speak at many meetings throughout the state in the summer and fall of 1995 in order to give my perspectives as an insider to the industry and an outsider to industrial privilege. My telephone rang day and night as growing numbers of Iowans were facing these same problems. Not only were Iowans calling, but Pam and I began receiving telephone calls from all across the United States. Other leaders from Friends of Rural America and I began doing interviews with radio stations, television news stations, newspapers, agricultural publications, and environmental magazines. As these broadcasts and publications were circulated, more people became familiar with our issues and were looking for help. What started as a very personal issue had become a large and looming issue all across the nation.

During the summer of 1995, several of us from FRA set up booths at county fairs in Iowa. We passed out information, talked with citizens, and had petitions signed. Everywhere we went we found people concerned about the future of independent hog production for themselves and their children. We found that people were concerned about the impact of large-scale hog operations on their property values and their health. They were concerned about toxic build-up from abandoned sites and contaminated aquifers and wells from leaking hog manure lagoons. They were concerned with the overapplication of nutrients on the land from the large volumes of hog manure. They were concerned about the high clean-up costs of abandoned sites for Iowa taxpayers. They were concerned with the impact of changing demographics on their small businesses and schools as traditional farm owners were being replaced by more transient employees who were less likely to own property and sit on church and school boards.

The concerns Iowans had over these developments were too numerous to be ignored by our state legislature. During the summer of 1995, we read almost weekly about disasters occurring in North Carolina, Iowa, and Missouri, such as water contaminations and fish kills due to large-scale hog operations. We who were experienced in hog production predicted there would be manure spills galore. There were. We said there would be a massive expansion of industrial sites built by absentee owners. There were. We said that Iowans would not be content to have all local control stripped from them while they were to retain local control over all the social, economic, and environmental costs of such expansion. They weren't. We were told that the hog issue was dead. It wasn't.

I took all of these concerns and information to various state legislators who were looking for real solutions. I found that many representatives from both parties in both the Senate and the House were willing to work together. When the 1996 legislative session began, I was able to build a network of legislators dedicated to seeking remedies to the problems of industrialized hog production. As the new lobbyist for FRA, I helped draft three major bills that would have helped family farmers and would have slowed industrial expansion.

The first bill dealt with price discrimination at the packing plants; the second was a packer-feeding bill; and the third would have banned earthen lagoons as storage pits for manure. The first bill passed the Senate with overwhelming support on a vote of 48-1. The packer-feeding bill passed out of committee but was never brought to full vote on the Senate floor due to lobbying efforts from the Iowa Pork Producers' Association. During the negotiations of this bill, the

IPPA lobbyists set about undermining most of the stipulations that represented the core of the bill. We could not endorse a bill that prohibited packer-feeding in name only and left all family farmers as vulnerable as before. The third bill on lagoons was defeated in the Senate. A full year of intensive lobbying efforts was defeated. It was clear that big changes were needed in the next election in order to get anything accomplished.

With critical elections near, we followed the campaigns with intense interest. Various candidates did what they could to raise hog concerns and win their elections. The election results were disappointing. Still, with less hope of getting any significant legislation passed to help regulate industrial hog farms and protect traditional hog farmers, we continue to be diligent. The 1997 legislative session contended with concerns over Iowa Geological Survey reports concerning groundwater contamination. Groundwater contamination by hog waste in any one area was now believed to be capable of contaminating vast areas of our interconnected underground water supply. Since many of the earthen manure lagoons in our region are currently constructed below water table levels, the survey revealed a grave danger to the water supply for most of Iowa. Even this danger was not enough to cause most of our state policy makers to question their support for large-scale hog operations.

INDUSTRIAL HOG MYTHS DEMYSTIFIED

Much rhetoric is distributed today which incorrectly lauds the benefits of industrialized pork. From outside the industry and the communities being affected, this rhetoric seems very convincing. For many within the industry and within the communities being affected, the rhetoric is hollow. It reflects a lack of understanding of the real inner workings of the hog industry or a vision for it being socially acceptable, economically beneficial, or environmentally sustainable. These three elements have allowed Iowa to raise 25 percent of the hogs in the nation for the last 120 years without negative front-page headlines and continual neighborhood conflict. The loss of these elements is causing much of the controversy surrounding this very old, very large, and in the past, very well accepted industry in our state.

State policy makers in the early 1990s formulated a plan to increase Iowa's national share of meat production from 1 to 5 percent. They could have chosen to support and expand traditional Iowa methods of production, but instead they chose to promote this increase through industrialized production methods patterned after the model developed in North Carolina that Senator Robert Morgan describes in his chapter. Since then, volumes of misinformation lauding the efficiencies of large producers and degrading traditional styles of production spread to farmers through farm and land grant university publications. Believing this misinformation, thousands of Iowa hog farmers liquidated their herds. Unfair and illegal pricing structures which subsidize industrial producers at the expense of independent farmers were allowed to be developed by packing plants. Lower interest rates from lending institutions, federal and state tax advantages, and property tax revenues being used to educate and pay industrial employee wages all help to prop up the industrial hog expansion as Blaine Nickles discusses in his chapter. Iowa had now marked the cards and stacked the deck in favor of the industrial hog producer. Hard work, efficiency, and cost of production took a back seat to "relationship" and political advantage in determining who would be Iowa's future pork producers.

The current controversy in Iowa is not over hogs. The primary controversy is over the structure of the industry. We have had many more hogs in Iowa in the past than we have today, and they were raised without much conflict. Iowans know how to raise hogs in ways that are socially acceptable and environmentally sustainable. The industrial model of pork production is neither of these. The pork industry is going backward with its promotion of outdated industrial models. Thirty million gallons of raw, untreated hog manure stored in open earthen holes in the ground is not new technology. Neither is it socially acceptable. Iowa is allowing what benefitted it for 120 years to be destroyed while at the same time selling itself out to a cheap industrial imitation. Common sense and social concern have left the industry. Blind greed has taken their place.

The discrepancies between commonsense social concerns and the current trend toward blind greed can be illustrated in many ways. First, in our community we care about the safety of our drinking water and the condition of our land from which we make our livelihoods. We value our ability to enjoy backyard barbecues, hang our laundry out on the clothesline, and take evening walks for exercise in the fresh air. If a neighbor's farm practices threaten these, we drive to their homes and talk with them to find a solution. When the farmer lives on the farm in the community, solutions can be reached.

We deeply value certain social benefits long extolled in our rural communities. We value the low crime rate that allows our children to play outdoors without fear of rape, kidnapping, being shot, or encounters with drug dealers. We watch out for each others' children and homes and console each other in times of tragedy. If we break something that belongs to a community member, we make restitution. We value the time spent at our golf course, local restaurant, and cafe where we socialize. We commit time and financial resources to see to it that our children and elderly are well cared for, that our churches serve our congregations thoughtfully, and that our schools provide the best opportunities and educational environments that our communities can provide.

Much of our business is still done with a handshake and the good-hearted comment, "I know you are good for it, and if I have any problems, I know where to find you." This community trust and sense of mutual responsibility is part of the reason why the independent hog producer networks Randy Ziegenhorn describes in his chapter later in this book are successful. We support our local business people because we watched as they baptized their babies, cheered for their children at ballgames, sat beside them on community boards, all for the purpose of making lives together. Pork producers and pastors, grocers and mechanics, secretaries, teachers, and homemakers alike have taken pride in our families, homes, and community life.

In contrast, the blind greed that drives the structural changes of the hog industry threatens all of the community values that we cherish. Earthen storage basins and open manure lagoons are leaking and contaminating wells and other waters. Improper applications of manure threaten to saturate our land with overloads of nitrogen, phosphorous, potassium, copper, and zinc. Some rural residents can't even open the windows of their homes on sunny days because of the pungent odors and gases from open manure pits. Owners of offending facilities frequently are inaccessible or oblivious to these complaints from frustrated neighbors since these owners often live elsewhere rather than on the site of the hog units. These owners are far removed from any positive social peer pressure to protect local environments. Neighbors resent the lack of concern for the decline in their property values, health, or quality of life they enjoyed for generations. Lines are drawn between lifelong friends either for or against these facilities, and this erodes the sense of neighborliness in Iowa communities similar to what Laura DeLind reports in Michigan. And when environmental violations occur, passions run high as government officials fail to enforce laws.

Social life changes as traditional rural families leave to find new jobs elsewhere and are replaced by underpaid transient laborers working in the large-scale facilities. What Mark Grey describes in his chapter concerning the community effects of the packing industry is now occurring on the production end. Newcomers tend to be less involved in local community life through churches and school. Seldom do they own properties that help to ensure proper upkeep of a community. Churches and schools consider consolidation because of the decline in rural populations, and the increasing social burdens that accumulate add to the sense of burden. Local businesses find that if they have not been bypassed altogether by aggressive owners "looking out for number one and freed from subsidizing smaller businesses," then it is often difficult to collect payment for the services they provided. Some corporately run hog producers have left businesses behind with only partial payments for their services and products. Rural communities, residents, and area resources are strip-mined and left buried by the residual effects of blind greed.

I laugh every time I hear the statement that this antiquated industrialized method of hog production is clean, free from disease, and economically efficient. I have raised hogs all my life and raised them in confinement since 1970. There are advantages and disadvantages to all methods of pork production.

Iowa needs honest research which will give accurate and unbiased information to citizens concerning the advantages and disadvantages of all types of production. The antiquated technologies being promoted by the ruling structures in Iowa and North Carolina are not the panacea we are being led to believe.

Iowans need to know some of the practices common within the walls of industrial production units. What people are not being told is that many so-called high producing sows are having difficulty conceiving. To produce large groups of pigs the same age, some of these "high-tech" operations breed young females and then group-inject an abortion drug so these young females can be rebred and have babies en-mass. Because some high-tech breeds of sows do not cycle normally, they are given a shot to help them cycle. Then they are given a shot to induce labor. When they are weaned, they are given another shot forcing them to re-cycle. After two or three farrowings the sows are so worn out that they are sold for slaughter or dragged from the farrowing crate and killed because they are unable to walk to the slaughter house.

Do these large units provide stable employment and community benefits? Perhaps it would be helpful to look at Dows, Iowa, near our farm. Dows is a beautiful community located near the heart of mega-hog expansion in Iowa. In a recent discussion with several of Dows' schoolboard members, concern was expressed over the number of transient students in their school district. Teachers find it difficult to know upon which students to focus their attention. Should it be the established students who progress steadily, or the transient students who are far behind in their studies? Because enrollment continues to decline, whole grade sharing or consolidation is unavoidable. The additional instability brought by these industrial units is affecting the very fabric of our schools.

The Northeast Hamilton School District lies next to the Dows Community School District. This district is also located in the heart of mega-hog expansion. The Iowa *Falls Times-Citizen* reported (Feburary 21, 1996), that falling property values would affect the district in the next year. Gary Schnellert, superintendent of Northeast Hamilton, said that he was:

> particularly concerned about a $6 million drop in ag land valuations, and the nearly $1 million in Iowa Select Farms property taxes that the district will not be receiving. Iowa Select has received a new jobs training program agreement that will funnel property taxes, including the school district levy, into job training for the company's employees. But we still have to educate those employees' children. Other taxpayers will have to foot the bill.

In my own community of Latimer, one can barely find business owners who have not been adversely affected by one industrial-style producer in the area. This producer does not use local concrete companies to pour his facility foundations. He built his own concrete plant. He brings in outside contractors to build his facilities rather than hire local contractors. He does not hire local truckers to move his products. He bought his own truck line. He does not buy his hog feed from the local feed salesman. He built his own feed mill. Local businesses have found it difficult to collect full payment from him for their services and refuse to do further business. Grain prices have not risen for local farmers due to increased livestock production, despite what economists predicted should happen in free market systems. Free food pantries have been depleted. And local officials have lamented the increases in crime and the heavier work loads in the court system.

An employee of a large-scale corporate producer recently rented a house in town that is owned by my father. The family fled in the middle of the night, leaving two junk cars and their belongings behind. My father wonders how to handle the situation and his damaged rental property. At least the family took their car's engine off of the chain that hung for two months from the mulberry tree before they skipped town. Claims concerning rural economic development have little credibility among community members affected by this industrial-style production.

We are being told that consumers are demanding high quality pork that is consistently lean. Unfortunately, the pork being produced by the industrial units meets the packing plants' qualifications for lean pork but does not meet the consumers' desire for high quality meat. A two-part study by the National Pork Producers Council released in April 1995 revealed that consumers are not demanding the type of pork being promoted by the pork industry today. The study revealed that consumers prefer pork with more lipid (intermuscular fat). It is this lipid that adds to the full flavor and tenderness of the pork. Consumers would probably eat more pork if it were not so tough. The packers, not the consumers, are demanding lean pork. Dennis DePietre (1996:64) from the University of Missouri's Commercial Swine Division writes,

> A buying system which pays on the basis of weight and lean percent is a low knowledge system. It is a low knowledge system because it is not accurately pricing the characteristics demanded by consumers. In the absence of knowledge of real consumer demand, it is pricing pigs to lower the cost of slaughter and carcass breakdown. It is a pricing system based on packer technology rather than consumer preferences.

Consumer demand is not driving the hog industry today. The type of pork demanded by consumers at home and abroad is not being raised by the industrial producers, it is being raised by family farmers.

Unfortunately, it appears that the independent producers' high quality hogs are being purchased for less at the packing plants because they are being told those hogs carry too much fat. Because these hogs produce high quality meat, it is then sold at a premium price both overseas and at home. The poorer quality meat being produced by the industrial producers exacts a premium price at the packing plant, but then it is sold at discounts or is made into sausage because much of it is of such low quality.

The meat consumers are demanding can be raised in Iowa on our traditional hog farms. This is a fact that the industrial producers do not want us to know. Traditional Iowa producers are raising the quality of meat and have the skills necessary to make pork the meat of choice worldwide by the year 2000. All they need is accurate information and a level field on which to compete economically. Rather than defending illegal payment practices at the packing plants and promoting misinformation within the industry, the leaders of the pork industry, our legislators, and our agricultural universities need to begin to dispense accurate information and stand up for justice and fairness for all producers. Because justice and fairness are no longer a part of the pork industry, "access to capital" is now denied to anyone who is not in a proper "relationship." The creation of a pork production system that promotes low cost, high efficiency producers who raise hogs by socially acceptable and environmentally sustainable methods should be our goal.

CONCLUSION

More and more family farmers fear financial failure and enter into contract production agreements ("relationships") with industrial-style hog companies. Farmers are receiving promises of risk-shared pork production where they can eventually own their own facility. Through this process each farmer becomes little more than an indentured servant. If I were to enter into such a contract, I would lose control of my breeding stock, my feed quality, the health care of my herd, and my profits. I could not even control whether to keep my buildings full of hogs in order to keep making payments on the buildings I would hope to own someday. As is already apparent in the poultry industry described by John Morrison in his chapter, the large integrators would possess all power over my ability to keep my buildings full to pay my bills. I would receive the responsibility for the labor, utilities, manure management, and productivity of the herd without having the tools to do my job to the best of my ability. All of this responsibility provides less income for my family than fair and open markets have provided in the past. Such a reversal in farm economics across the whole rural community translates into financial reversals for every business we farmers have frequented in the past. This is not an option for me and my community.

It is not an option for me because of what it will do to my community of friends and farmers and to agriculture and hog production as a whole. Economic benefits would eventually go to those who bear little concern for our community, and little of the life and values we cherish would remain for our children. Without profits for our farm products, eventually even the land will be owned by large corporate investors from our country and other countries involved in transnational corporations. Local economies will crumble, neighbors will lose access to local people who can address area concerns, and local control over the defining structures of our lives will be lost. Our ancestors understood the importance of independent farming for our country's well-being.

To remain in production in the same manner as I have for the past twenty years leaves me vulnerable to the packers and their "honor system." An article from a newspaper in North Carolina (Wilmington Morningstar, August 4, 1996) this past summer reflected my concern for fair and open markets. A spokesman for Smithfield Foods, a packing plant in Tar Heel, North Carolina, addressed the issue of market access within his state. Due to the increase of contracted hogs within the state, this spokesman said that his company needed to increase the size of its packing plant. He stated that his plant is legally bound to honor its contracts of purchase agreements with the industrialized producers before offering to purchase hogs from independent producers. As the number of contracted hogs increases in a region, the number of independent hogs needed to satisfy the market place will decrease. When Pam discussed this issue with the vice-president of the Iowa Farm Bureau, he said that could never happen in Iowa. His naivete and wishful thinking alone will do me and other family farmers little good. I realize that I will likely find it increasingly difficult to find a meat packing plant that will purchase my hogs at fair prices. I must consider other options.

One option is to sell off my farm and all of my hogs. I have watched many of my friends do that over the years. I have helped at many farm sales and stood by my friends as they watched their machinery and equipment go to the highest bidder. These farmers have been grateful just to break even at the end of their day, only to realize at a later date that their tax liabilities have bankrupted them. This is not an option that I find reasonable.

Contract feeding is not an option I want to consider. Future markets remain unsure, and I do not want to sell out. Therefore, one of the options I am currently considering is to raise feeder pigs for other producers in an arrangement similar to the networking arrangements Randy Ziegenhorn describes in his chapter. I must be able to retain the employees I hire and stay profitable for my family as well. I am also closely watching developments in the growth of alternative food systems, such as Community Supported Agriculture programs (CSAs), truly independent co-ops, and stores and restaurants that are looking to purchase meats and that are respectful of family farmers and rural economies.

My family stands here in the midst of change and trusts that there will be a way for us as family farmers to continue to do what we do best. We want to raise our food and our families in the best way possible and to make our way in this world in a socially, economically, and environmentally positive manner. I remain as president of and lobbyist for Friends of Rural America. I continue to work on bipartisan efforts to help bring reasonable regulation to the expanding industrialization of hog production and to preserve a place of profitability for family farmers such as myself. The job gets harder every day, but the ineptitude of the industrial expansion may catch up with its promoters. The numbers of citizens writing letters to newspapers is growing continually, and the content of those letters becomes more and more educational as leaders from all cross-sections of life and science are emerging. We continue to learn from each other.

My view from inside the hog industry has given me the unique opportunity to explore all of the data and debate the conflict generates. It is gravely clear to me that unless public sentiment becomes public activism which then leads to public policy, little will remain within agriculture, farm communities, and eventually the entire democratic process for our children to inherit. It is not too late to recover the American dream of building a small business in agriculture if we defend what our forefathers came to this country to attain. In the effort to rebuild this dream and make it sure for those that follow, we should remember the words of Franklin D. Roosevelt (Davidoff 1952:60): "We build and defend not for our generation alone. We defend the foundations laid by our fathers. We build a life for generations yet unborn. We defend and we build a way of life, not for America alone, but for all mankind."

REFERENCES

Davidoff, Henry (editor). 1952. *The Pocket Book of Quotations.* New York: Simon and Schuster.

DiPietre, Dennis. 1996. Access to Markets. Columbia: University of Missouri.

Graham, Larry. 1995. National Pork Producers Council Memorandum, June 26. Des Moines, Iowa: NPPC.

Wilmington Morningstar. August 4, 1996. Hog Forecast: Expect More. Wilmington, North Carolina.

Chapter 3

Meatpacking in Storm Lake, Iowa: A Community in Transition

Mark A. Grey

INTRODUCTION

Storm Lake, Iowa, reflects how rural communities in the United States are struggling with changes in the meatpacking industry. It was here in 1982 that Iowa Beef Processors (IBP) opened its first major pork packing plant to revolutionize pork processing in the same way it revolutionized beef. Meatpacking moved from large urban centers to rural locations to access large animal inventories. The attraction of rural Storm Lake for IBP is that it is centrally located in a highly productive hog region with inexpensive corn supplies (the primary feed for hogs) and the availability of an old plant for cheap renovation.

The availability of sufficient hogs produced by Iowa farmers to keep IBP in Storm Lake has been called into question. The president of IBP charged that Iowa is biased against big business and that the state must change its hog production policies in order to remain attractive to major packers (Wagner 1994). This charge was followed by an announcement that the corporation is set to establish a site for a new "mega-plant" capable of slaughtering thirty thousand head per day somewhere in the southeastern United States.

Assessing whether rural communities in Iowa and elsewhere would suffer from an IBP departure assumes the plant's presence is beneficial. Storm Lake is one of several midwestern communities experiencing the consequences of a new breed of meatpacking plant. Each community must decide for itself whether the costs of these new plants outweigh the benefits. Given the central role of hogs and pork production in Iowa's economy, most economists would argue that the presence of large, efficient plants helps sustain our agricultural economy. However, the economic benefits measured in gross dollars and number of jobs contrast with the community consequences and experiences of community members who believe the costs are too high.

Storm Lake reflects the transformation of meatpacking labor with attendant social changes in the community. Most Storm Lakers, as established residents prefer to be called, told me "things have never been the same" since IBP came to town. More specifically, things have not been the same since IBP's workers arrived. Most Storm Lakers involved in my research determined that the community costs of IBP outweigh the benefits. These costs include rising crime rates and increased stress on community institutions such as schools and health care providers. Both of these themes are placed within the context of Storm Lake's meatpacking history (Davis 1982–83; Grey 1995a).

This research was conducted between 1991 and 1994, when I interviewed approximately one hundred people in the community. I selected most informants for their specific roles in the community and recent arrival or experience in the meatpacking industry. I used translators during interviews with Lao and Mennonite informants who could not speak English. I also used secondary sources such as newspapers and company documents.

MEATPACKING IN STORM LAKE

Storm Lake's long association with meatpacking began in 1935 when Storm Lake Packing Company was opened by Kingan and Company of Indianapolis, Indiana, and Nash Brothers. Kingan bought out the Nash Brothers two years later. As one of Kingan's six slaughter facilities, the plant initially processed cattle, hogs, and sheep (Buena Vista County Historical and Genealogical Society 1984). Its daily capacity grew from 500 to 2,100 carcasses in 1947 after Kingan expanded the plant's cooling facilities, power plant, and cut floor. By 1950, the plant employed some 370 workers.

In 1952, Hygrade Food Products Corporation purchased control of Kingan and Company. With the addition of Kingan's six major plants, Hygrade became the nation's fourth largest meatpacker. In October 1973, Hygrade increased the plant's capacity by 60 percent and the workforce grew to about 650 (Grey 1995b).

The Hygrade plant was highly productive. Former workers boasted that during the final years of operation, the plant produced more pork per worker than any other plant in the nation. Of all sows slaughtered in the United States 18 to 20 percent were being processed in Storm Lake. In addition to three thousand sows per day, four thousand butcher hogs were regularly slaughtered (Davis 1982–83).

Storm Lake became a union meatpacking town in 1937 when Local 191 of the Amalgamated Meat Cutters and Butcher Workmen of North America (AMCBW) was formed. In 1978, Local 191 joined the United Food and Commercial workers (UFCW) and retained this affiliation even after the plant's closure in 1981. Former members noted that all line (non-management) workers were enrolled in the local and new workers were encouraged to sign on by plant managers (Davis 1982–83; Grey 1995b).

The Storm Lake plant was marked by a lack of conflict between workers and managers until its final years of operation. The local union threatened to strike only once. This was in 1976 when the corporation delayed signing a new contract. No wild cat strikes were called during the forty-four years of union representation. In February 1981, Hygrade announced that the Storm Lake plant would shut down because of rising energy costs and "unfavorable" labor rates.

Hygrade's closure was a significant turn of events for the community. When the plant shut down, some six hundred high-wage jobs were lost, as well as Buena Vista County's principal consumer of finished hogs. At the time, community leaders seemed philosophic—Storm Lake would "survive." Besides, the Hygrade facility was "too good" to sit idle. They were convinced a new owner would come along, and the city began soliciting potential buyers. Business leaders seemed most concerned about the loss of Hygrade's $15 million annual payroll. But these concerns were tempered by their contention that wages at Hygrade forced higher wages among other Storm Lake employers. Indeed, some suggested that Hygrade's high wage structure (1981 average of $10.69/hour plus incentive pay) was actually a deterrent to attracting new industry.

Hog growers found less competition for their hogs and had to find new markets. The closest pork plant was the Wilson Foods plant in Cherokee. Other plants were fifty to seventy-five miles away. Local pork producers expected lower prices and higher transportation costs, both cutting into their profits (Harris 1981).

IBP COMES TO STORM LAKE

In 1979, IBP's president, Robert Peterson, announced the company's intention to enter pork production. With the purchase and renovation of the Hygrade plant in Storm Lake, the company's pork division vice president said, "Our dream has come to reality" (Herron 1982b:1). IBP purchased the old plant for $2.5 million, and then renovated it at a cost that exceeded the $9.7 million in industrial revenue bonds issued by Storm Lake as an incentive to buy the plant.

Its opening five months later was announced with considerable fanfare. Iowa's Governor Ray was flown to Storm Lake on IBP's corporate airplane. He was joined by the company's president and a host of other dignitaries for an opening ceremony, plant tour, and invitation-only luncheon. The Storm Lake High School marching band played, and red carpets were rolled out (Herron 1982a; Herron 1982b). Iowa's governor cited IBP's reputation for "being aggressive, innovative, creative and being revolutionary. It has done all of this with the beef industry and everyone expects them to do it with the pork industry" (Herron 1982b).

Many in Storm Lake were excited that IBP had chosen their community to start the "pork revolution." A former president of the National Pork Producers Council called IBP's move "a giant step forward for our industry" (Herron 1982b). Others welcomed IBP because they believed the new plant would provide jobs for former Hygrade workers who had several years of experience in the industry, were already settled in the community, and were, for the most part, unemployed and readily available. But IBP had little interest in these workers. Several hundred applied, fully aware of the substantially lower wages, but fewer than thirty were hired. Some Hygrade workers believed IBP was concerned about their history of union activism (Grey 1995b). IBP did hire the former president and vice-president of the Storm Lake Local of the UFCW for management positions. However, IBP denied hiring them to undermine the union.

IBP used its new plant in Storm Lake to launch its efforts to take a larger share of pork packing. Within six years, IBP was the nation's largest pork packer. The corporation currently owns five pork plants in Iowa and one in Nebraska. Another is scheduled to open in Logansport, Indiana. IBP's daily slaughter capacity is 60,800 or 16 percent of national capacity. At present, the Storm Lake plant has the second largest capacity among IBP's pork plants, processing up to 13,400 head per day or 3.35 million per year. It is also the corporation's most profitable plant (Center for Rural Affairs 1994).

The transformation of pork processing meant the introduction of technological and other innovations that IBP perfected in its beef operations. These changes emphasized production efficiency and adding more value to livestock. For example, the IBP innovation of "boxed pork" was borrowed from beef processing plants (Tinstman and Peterson 1981), a process through which hog carcasses are "disassembled" to increasingly smaller pieces, shrink wrapped in the plant, and boxed for shipment directly to retail outlets. Cutting the hog into smaller pieces in the plant eliminated the need and expense of skilled butchers in grocery stores. Prior to boxed pork, it usually arrived at grocery outlets in "sides," which were then cut into consumer-sized pieces. Today's smaller pieces of pork often arrive already boned and need only be cut into individual pieces, a process that requires no particular skill beyond the safe use of a bandsaw.

Box pork means plants now add more value to hogs and therefore control a larger proportion of profits. Middle men are eliminated because the meat is shipped directly to stores, and shipping costs are reduced because more product can be carried in tightly stacked boxes on trucks.

Processing pork into smaller pieces requires more labor. However, as the case of Storm Lake illustrates, this did not always lead to recruitment of local workers. In contrast to their predecessors, the new breed IBP plants often avoided hiring experienced workers from local communities because of their previous affiliations with unions and expectations for high wages. Instead, new packers rely on labor sources outside their immediate communities and the state. These new workers are often minority refugees and immigrants. By 1992, IBP's Storm Lake workforce grew to 1,200, one-third of whom were immigrant or refugee minorities.

The import of labor was accompanied by downward pressure on wages. Between 1980 and 1991, hourly wages in meatpacking dropped from $10.75 per hour to $5.65 per hour (adjusted for inflation) (Kasler and Steimel 1992:1A). IBP reopened the Storm Lake plant in September 1982 with relatively low wages ($6.00 per hour to start). A few years later, they were $6.50 per hour and remain so today. By 1993 the workforce grew to over 1,300 and the annual payroll was $27.5 million, nearly double Hygrade's 1981 payroll of $15 million. When adjusted for inflation, however, today's IBP workers earn about one-third of the income of their counterparts at Hygrade.

IBP also opened a number of new plants without unionized workforces or where unions were effectively neutralized. Today, only four of IBP's seventeen plants are unionized. The Storm Lake plant was opened without a union and remains so today.

STORM LAKE'S NEW WORKFORCE

For the first few years, IBP found sufficient job applicants in northwestern Iowa. But its low wages, notoriously high turnover rates, and a growing overall workforce meant that a number of jobs would go to newcomers. By the mid-1980s, Southeast Asian refugees, mostly ethnic lowland Lao, began arriving in Storm Lake from other states and Iowa communities. Often they are referred to as "valley" Lao because of their residence in the Mekong River valley in Laos. Although some twenty-four Tai Dam refugee families were sponsored by Storm Lake churches and families to settle in Storm Lake in the 1970s, most of these refugees left the community within a few years. The Tai Dam are also known as the "Black Tai" and hail from the highland regions of the Laos-Vietnam border. Highland Tai Dam and lowland Lao are culturally very different. The newcomer Lao in Storm Lake were different from their Tai Dam successors in important ways. For example, the Lao generally migrated from the place of their original placement by refugee agencies, making them secondary migrants. They were attracted by jobs at IBP and growing kinship networks, which provided valuable assistance in migration (Grey 1993).

For many of the refugees, the availability of jobs at IBP alone was not sufficient to risk moving to Storm Lake. The added incentive was becoming part of a network of fellow Lao who provided the support necessary to assure a successful move. But their status as secondary migrants among Storm Lakers contrasted with the status of Tai Dam refugees. The Tai Dam were, in a sense, "invited" by the community; the Lao were not.

The new refugees differed in other ways as well. Most adults did not speak English. But the Tai Dam, who had already developed English skills, worked daily with Storm Lakers who considered the acquisition of English a sign of assimilation. Most Lao were Buddhists in contrast to many of the Tai Dam who were or became Christians and joined local churches. But the most distinguishing characteristic of the Lao was their relatively large immigrant population. By 1991, The Iowa Bureau of Refugee Services counted 149 refugee households, or 504 individuals in Storm Lake, and the number has continued to grow. Among these families, 125 household heads worked at IBP. In 1992, more than 300 Lao worked at IBP, nearly one-fourth of its workforce (Grey 1995a).

Other newcomers to Storm Lake included Mexican Mennonites. About seventy migrated to Storm Lake from Mexico and Texas in order for the male heads of households to take jobs at IBP. The diversity of languages, citizenships, nationalities, and immigration statuses among these people were remarkable. The men were directly recruited by IBP and its agent at their settlements in Chihuahua Province. With job offers in-hand and the assistance of IBP's agent (whom the workers also paid), they easily obtained "green cards" to legally enter the United States. But, for the most part, their families entered the country without documentation. Exceptions were children born in the United States who automatically gained U.S. citizenship.

Mennonite and Lao workers shared jobs available for non-English speakers at IBP. But an important distinction should be made in terms of how they arrived. IBP hired a Texas-based recruiter who was familiar with the Mennonite community and who spoke German to approach men in Mexico, provide information about jobs in Storm Lake, and assist in obtaining necessary immigration papers.

None of the Lao I met were directly recruited by IBP. Instead, IBP was able to tap into and redirect existing migration flows sustained by kinship networks and patron relations. The Lao workers came to the plant, rendering recruiting efforts unnecessary. The plant also hired a Lao-speaking Tai Dam man as a personnel director. His presence in the plant provided more Lao with jobs, but they became so dependent on him that his power and status elevated him to the role of patron. He continues to broker relations between newcomer Lao who do not speak English, plant managers, and the local English-speaking community (Grey 1993).

While the arrival of these two ethnic groups is significant, it is equally significant that until about 1992, the number of Latino newcomers to Storm Lake was minimal. Most migrant workers in other meatpacking communities are Latino, but in 1992, only about eighty Latinos worked at IBP (Grey 1995a). After IBP stepped up efforts to recruit Latinos in Texas and southern California, more Latinos moved to Storm Lake. Although I have heard estimates of more than three hundred, I am not aware of any accurate counts.

Latino newcomers followed different migration patterns than the Lao or the Mennonites. Most arrived without families, and because they are less rooted (and perhaps less welcome) in the community, they are more itinerant. In addition, the influx of Latinos had a greater psychological consequence than the earlier Lao migrations. Criticisms from community members reflect contrasting stereotypes about Mexicans and Asians: Mexicans are "lazy" and "don't stick around" because they don't like the work; Asians arrive with their families, "work hard," "pay their bills," and "are smarter." Variations on this theme were heard often, even among important service providers.

What is important to many Storm Lakers is that the arrival of Latinos signaled that the costs of IBP's presence were becoming too high. Of course, the presence of Latinos alone does not explain this view. It simply means that their arrival coincided with a number of other changes.

COSTS AND CONSEQUENCES OF THE NEW WORKFORCE

Schools

Newcomers often enroll their children in school even before they take jobs or find housing. In Storm Lake, the school system was confronted with challenges associated with a new population of students speaking little or no English. In addition, there were cultural differences to contend with, as well as a lack of records, appropriate testing devices, and no assurance that newcomers would enroll for extended periods. Indeed, one of the worst problems in this, or any other meatpacking community, is that high employee turnover disrupts schools and their curriculum.

When special students are enrolled, the school district is obliged to meet their educational needs. The costs can be high since often new teachers and aides must be hired to provide training in English as a Second Language (ESL). As a result, the newcomer children cost more to service than English speakers. The number of non-English students in Storm Lake schools grew rapidly from 28 in 1982 and 1983 to 236 in 1992 and 1993 (Grey 1995a). By 1995, 384 or

21 percent of Storm Lake students were eligible for ESL programs. Today, invitations to parents to enroll their children in kindergarten are published in three languages: Lao, Spanish, and English (*Storm Lake Times,* October 11, 1995)

In 1992 and 1993, Storm Lake schools received $157,000 in federal funds for ESL instruction, but those funds were being phased out. By 1994 and 1995, the ESL budget was down to $127,575, and federal contributions were scheduled to end. The district found itself in the difficult position of either paying more of these costs from their own budgets, laying off ESL staff, or both. There was resistance to shifting funds from mainstream programs, but the school board voted to use reserve money to fund the program for $100,000 in 1995 and 1996. Some board members proposed funding only half the costs and asking IBP and the Bil-Mar turkey plant (which also uses immigrant labor) to come up with the other $50,000 (*Storm Lake Pilot-Tribune,* October 25, 1994). Although the board did not pass this proposal, it was very unlikely that IBP would have acquiesced anyway. It would have been an admission of responsibility for the presence of newcomers and their children, an admission the corporation takes pains to avoid.

Health Care

The health care sector was also hard hit. Newcomers brought new health problems to the community. In 1994, the county health department conducted tuberculosis (TB) tests in Storm Lake schools for the first time. In the high school's ESL program, 42 percent of students tested positive. Preventative medication was administered to more than fifty students, all refugees and immigrants.

Most problems for the clinics and hospital are associated with language barriers and employee turnover at IBP. Turnover in particular has led to tremendous growth in uncompensated care. At Storm Lake's largest clinic, which provides services for most IBP workers and their families, minority patient numbers grew from a total of 15 nearly ten years ago, to more than 20 percent of all patients in 1994.

About ten new languages were brought to the community, and communication became a major barrier to adequate health care. Except for one physician who speaks German, there are no personnel fluent in these languages. Patients must bring translators with them. These translators are often children, and the language needed for diagnosis and treatment goes beyond their limited vocabularies. As a result, the doctor's diagnosis may take up to three times longer, leaving less time for other patients. This is critical in light of the growing number of patient encounters. With the same number of physicians, patient encounters increased 36 percent from 42,000 in 1984 to 57,000 in 1993. According to the director of the Buena Vista County Health Department, even if qualified translators could be found, hiring them would be prohibitively expensive.

The community's largest clinic has seen uncompensated care expenses grow. Prior to the mid-1980s, uncompensated costs were minimal. Today, the clinic has more than $10,000 a month in unpaid services. This excludes "write offs" associated with Medicare and Medicaid.

Buena Vista County Hospital has also seen its patient numbers grow with 35,000 total encounters in 1993. Of these, a growing percentage were non-English-speaking minorities presenting the same language barriers experienced at the clinic. Unlike the private clinic, the hospital is funded by Buena Vista County taxpayers and may not turn away patients for any reason, including lack of payment. One result was the doubling of emergency room visits in the last ten years to seven hundred to eight hundred per month. An increasing percentage of emergency visits are made by newcomers who use the service as a provider of last resort.

Unpaid bills are approaching crisis levels for Buena Vista County Hospital. Uncompensated care is now the fastest growing item in the hospital budget. While some of this is "contractual write-off" associated with Medicaid and Medicare, an increasing proportion is simply unpaid bills from newcomers. Ten years ago, total uncompensated care amounted to $40 to 50 thousand per year, or only about 1 percent of the budget. In 1993, the hospital wrote off $1 million, or 13 percent of its total budget. As the hospital administrator put it, if uncompensated costs reach 15 percent, "arms and legs are going to start coming off around here."

A critical factor in uncompensated care at the clinic and hospital is the IBP health insurance package. Health benefits are available only after six months on the job. Any health related expense during the first six months on the job is the complete responsibility of the patient. Because of worker turnover, many employees do not work six months, so their bills go unpaid.

Crime

Nothing has driven home the concerns of Storm Lakers more than rising crime rates and a growing sense that Storm Lake is not the safe place it used to be. One resident assured me that Storm Lake now has the "highest number of police per capita in the state." The police department budget in 1994 and 1995 was $820,886, a 35 percent increase over the 1990 and 1991 budget.

In 1994, the 599 serious crimes reported in Storm Lake were two and a half times greater than those reported in other comparably sized Iowa cities. In 1994, the police department received 39,191 calls (average of 109 per day), up 10,000 from 1990. Between 1990 and 1993, calls for fights and disturbances grew from 59 to 272. Burglary calls tripled to 145 in the same period. The number of inmates in the Buena Vista County Jail swelled. In 1992 the total number of inmates was 804, 63 more than in 1991 and 211 more than 1990 (Grey 1995a).

Beyond the numbers is the perception that Storm Lake is no longer a safe place to live. Many of these perceptions are supported by statistics, but the sense of insecurity is also driven by high-profile incidents, many involving firearms. In one prominent case, four Lao men were arrested for possessing stolen property, transporting a semi-automatic pistol without a permit, and possessing a Chinese-made assault rifle (Grey 1995a).

The police department responded by hiring more patrol and community service officers to work directly with the Latino and Southeast Asian communities (*Storm Lake Pilot-Tribune,* June 11, 1994). One of these officers is fluent in Spanish, and another in Lao. Neither carries a firearm, and both act as liaisons between newcomers and the department. It is too early to tell if this program will affect crime rates.

Concerned that Storm Lake was getting an undeserved bad reputation, business leaders attempted to keep the police department from faxing press releases about major crimes to the statewide media. The president of the Chamber of Commerce argued that competition between Storm Lake's two newspapers sensationalized incidents and led to perceptions that crime in the community was rampant (Martin 1994). Efforts to address crime rates usually parallel a defense of IBP. "All rural communities are changing," goes the argument, "not just Storm Lake, so you can't blame IBP."

IBP's recruitment and its perceived link to crime became the dominant issue of the 1994 election for Buena Vista County attorney. In the Democratic primary, the sixteen-year incumbent was challenged by an attorney who made IBP's hiring practices a central campaign theme. The challenger accused IBP of "social pollution" by recruiting workers with criminal backgrounds. IBP responded that it did not go outside the community or recruit people with criminal records. "In fact, about the only place where we have been taking job applications in recent years is at our plant employment office" (*Storm Lake Times* June 4, 1994). Of course this does not mean that IBP is not recruiting outsiders; it simply means that applications are being taken at the plant. In addition, IBP also expressed concern about the crime rate but denied contributing to it. "When we screen job applicants, we ask if they ever have been convicted of a felony. If the answer is yes, we seek additional information before making a decision whether to hire them" (*Storm Lake Times,* June 4, 1994). Even if applicants have been convicted of felonies, or if they fail to disclose their records, they may be hired.

The challenger won the primary, but there was some disagreement about whether he won because of his anti-IBP rhetoric or the incumbent lost because he cracked down on illegal gambling operations at the Knights of Columbus and Elks Club. The Democratic candidate continued to attack IBP during the general election campaign, but eventually lost.

BENEFITS VERSUS COSTS OF LARGE PACKING PLANTS

The community and rural development consequences outlined here are directly associated with IBP's control and manipulation of its workforce. Storm Lake has been caught up in the new rules governing labor availability. While Hygrade was in operation, its unionized workers held a relatively strong position relative to management because in the local community they were the only workforce available. IBP, however, relies on newcomers who are relatively weak because they are not the only workforce available and they do not have any organized representation. IBP will simply dismiss them to import others.

The notion of "jobs" in this context is severely distorted. Former Hygrade workers point out with pride that they were "family men" who owned homes, bought cars and boats, were stable, and supported the local economy. Also they could afford it. On average, their earnings were three times those of current IBP workers. Workers with extensive experience, high hourly wages, and weekly incentive pay could earn more than $40,000 per year in the Hygrade plane (Grey 1995b).

The story of how these jobs were lost was repeated around the country as old plants closed during the industry's transformation. Despite evidence that wages alone were not the central downfall of old plants (Lauria and Fisher 1983), the movement toward lower wages and benefits helped convert meatpacking into an industry fraught with labor turnover. In the early years, plants experienced turnover rates of 30 percent per month or higher. Even after plants have operated for years, annual turnover rates of 72 to 96 percent continue (Stull 1994).

There is debate among academics and meatpacking officials about whether the industry benefits from and fosters high turnover rates or seeks to lower turnover and stabilize its workforce. From the latter perspective, the industry points to the high costs of training new workers. Every worker who remains employed, they state, saves hundreds of dollars in training costs. Upon closer inspection, this argument is suspect because these corporations are particularly adept in taking advantage of state job-training programs that defray training expenses and wages.

Typically, these are a form of tax-increment financing authorized by the state. Similar to what Blaine Nickles discusses in his chapter, contracts are made with community colleges which sell bonds to cover training and administrative expenses. The bonds are then repaid by a combination of diverted property taxes from the plant and payroll withholding taxes. The colleges keep their share. Through this arrangement, the state may reimburse the plant for up to one-half of new workers' wages for ninety days. New hires earn the entry-level wage, but half of that wage may be paid by taxpayers. When workers quit after the ninety-day period, their replacement will be cheaper because they will also be eligible for state training funds. High turnover means lower labor costs subsidized by state taxpayers. Since 1983, the plant received about $250,000 in worker training and retraining programs (IDED 1994). In short, stable workers cost more.

High turnover rates also reflect the hazardous nature of meatpacking. With more efficient machinery to move carcasses ("the chain"), higher capacities, and cutting the animal into smaller pieces, workers are also suffering from more repetitive motion problems, particularly carpal tunnel syndrome. For those who have been in meat plants and are mindful of high turnover and injury rates, there is an obvious question to ask corporate officials who claim they would like to bring turnover rates down: "Why not slow down 'the chain?'" The answer is that profits are dictated by the amount of boxed pork shipped out the back door rather than the well-being of workers. IBP's most profitable year was 1994, when the company earned more than $182 million (Perkins 1995). The factor most easily manipulated to extract as much wealth as possible from every animal is the labor force. Meatpacking labor has become "disposable" (Stull et al. 1993).

CONCLUSION

Will Storm Lake itself become disposable to IBP? Our current era of capitalism reflects a remarkable shift in the geographic mobility of labor and capital (Harvey 1989). Given recent threats to open plants elsewhere and close Iowa plants, IBP has some state officials worried. Or perhaps these are less threats than veiled maneuvers to extract further concessions from the state and community. In discussing any further concessions, the real cost of IBP to Storm Lake needs to be considered. There are no foreseeable changes in IBP's recruitment patterns, and the influx of newcomers in and out of the community will continue. Many of the costs will continue to rise, and agencies will be forced to limit programs for newcomers or shift funds from mainstream activities.

In their private moments, many Storm Lakers are angry about the changes that have occurred. Their concerns range from the loss of good-paying jobs in the old plant to feeling compelled to lock their houses when they go to the store. These conflicts reflect the dilemma faced by many rural midwestern communities: Is the price of "economic growth" associated with large meatpacking plants too high? This is a question Storm Lakers have asked themselves many times. All Storm Lakers I interviewed expressed personal concern about how the community has changed, and many agree that the costs of IBP's presence are too high.

REFERENCES

Buena Vista County Historical and Geneaological Society. 1984. The History of Buena Vista County. Storm Lake, Iowa: Buena Vista County Historical and Geneaological Society.

Center for Rural Affairs. 1984. *History of Buena Vista county.* Storm Lake, Iowa.

———. 1994. *Corporate farming update! Spotlight on pork.* Special Report. Walthill, Nebr.: Center for Rural Affairs.

Davis, Merle. Interviews with Merle Davis, Iowa Labor Oral History Project. Iowa City, Iowa: State Historical Society of Iowa, 1982–83.

Grey, Mark. 1993. The failure of Iowa's non-English speaking employee's law: A case study of patronage, kinship and migration in Storm Lake, Iowa. *High Plains Applied Anthropologist* 13: 32–46.

———. 1995a. "Pork, Poultry and Newcomers in Storm Lake, Iowa." In *Any Way You Cut It: Meat Processing and Small-Town America,* ed. Donald D. Stull, Michael Broadway, and David Griffith. Lawrence: University Press of Kansas.

———. 1995b. Turning the pork industry upside down: Storm Lake's Hygrade work force and the impact of the 1981 plant closure. *Annals of Iowa* 54: 1–16.

Harris, Dean, "Hygrade Closing Could Hurt Farmers," *Storm Lake Register,* 24 October 1981, 3.

Harvey, Davis. 1989. *The condition of post-modernity.* Oxford: Blackwell.

Herron, Steve, "IBP Opens Tuesday; Gov. Ray to Attend Special Ceremonies," *Storm Lake Register,* 25 September 1982a, 1.

———. "Giant Step Forward in Our Industry," *Storm Lake Pilot Tribune,* 29 September 1982b, 1.

Iowa Department of Economic Development (IDED). 1994. *Iowa new jobs training program, Iowa jobs training program: FY93 annual report.* Des Moines, Iowa.

Kasler, Dale, and Dirck Steimel, "Blue-Collar Iowa Strives to Cope as an Era Ends," *Des Moines Register,* 15 November 1992, 1A.

Lauria, Mickey, and Peter S. Fisher. 1983. *Plant closings in Iowa: Causes, consequences and legislative options.* Iowa City: University of Iowa Institute of Urban and Regional Research.

Martin, Jeff, "Some Say Competing Papers Sensationalize Crime in [Storm Lake] City," *Sioux City Journal,* 9 May 1994, A12.

Perkins, Jerry, "200 New Jobs for IBP Plant at Perry: The Expansion Will Allow Processing of Meat to be Sold to Japan," *Des Moines Register,* 20 January 1995, 10.

Storm Lake Pilot-Tribune, 25 October, 11 June 1994.

Storm Lake Times, 4 June 1994.

Storm Lake Times, 11 October 1995.

Stull, Donald D. 1994. Cattle cost money: Beefpacking's consequences for workers and communities. *High Plains Applied Anthropologist* 14: 65.

Stull, Donald D., David Griffith, Robert Hackenberg, and Lourdes Gouveia. 1993. Creating a disposable labor force. *Aspen Institute Quarterly* 5: 78–101.

Tinstman, Dale C., and Robert L. Peterson. 1981. *Iowa Beef Processors, Inc.: An entire industry revolutionized!* Princeton, N.J.: The Newcomen Society of North America.

Wagner, Jay P., "IBP Chief: Iowa Biased against Big Business," Des Moines Register, 9 December 1994, 1A, 3A.

Part II

The Environment

The three chapters in this section describe the environmental consequences of large-scale swine production for the health of workers, neighbors, and the general ecology—particularly water quality. Similar to social issues, environmental factors are treated as "externalities" in economic development assessments and models. This means they are not generally included as measures of the costs and benefits of doing business. However, as these contributions demonstrate, an economist's model cannot alter the environmental reality that farmers and rural residents face every day. Unfortunately, economic assessments devoid of environmental consideration all too often form the reality for policy makers.

Kelley Donham, a veterinarian with over twenty-five years' experience in rural health research, describes the environmental exposures and health hazards for those working inside hog confinement buildings. He first leads the reader through a typical work day, including exposures to dusts and gases, for a family hog producer husband and wife team. Donham then goes on to summarize twenty-five years' worth of research that clearly documents numerous health hazards to swine confinement workers, most notably chronic respiratory problems which occur in about one out of every four.

Susan Schiffman, a medical psychologist, and her colleagues then describe their research showing the psychological health consequences for neighbors exposed to the odor coming from large-scale swine operations. In a closely controlled research study, Schiffman and colleagues compared the psychological health of people living near large-scale hog confinement buildings with the psychological health of a control group of rural residents matched by gender, age, race, and years of education. They conclude that the neighbors of the large-scale swine operation were experiencing significant declines in emotional health. While some industry leaders and policy makers downplay odor as simply a matter of perception, Schiffman makes a compelling case that the impact of odor on humans may be the result of the physical interaction between odor molecules and various physiological pathways in the human body, such as neural connections between the brain's odor centers and the immune systems. An important lesson to be drawn from Schiffman's work is that odors ought to be considered a public health issue, not just an inconvenience.

In the final chapter of this section, Laura Jackson, an ecological biologist, describes the consequences of large-scale swine operations on a region's ecology with specific attention to water quality. Jackson describes known and potential water quality consequences of excess nitrogen, phosphorous, and heavy metals in swine waste that becomes highly concentrated in large-scale swine operations. In so doing, she points to the inadequacy of science when influenced by political forces to appropriately frame and conduct environmental research on large-scale swine operations. The consequence is a number of festering environmental problems that require immediate attention at several interconnected ecological levels and along a series of biological dimensions. Jackson offers several positive suggestions to address these issues.

Taken together, these chapters clearly demonstrate the need to account for environmental and human health in assessing industrialized swine production. As with many other human and environmental health problems, the costs are borne by the general tax paying public while the profits are assumed by a few.

Chapter 4

The Impact of Industrial Swine Production on Human Health

Kelley J. Donham

Agricultural-related death and injury rates are nearly five times that for all other occupations in the United States. They account for approximately 1,500 deaths and 140,000 disabling injuries annually (National Safety Council 1994). Excessive injuries, illnesses, and environmental health hazards result from agricultural exposures which pose unique health and safety problems. Wider environmental problems, such as the contamination of ground and surface waters, which Laura Jackson describes elsewhere in this book, raise long-term health questions.

Agricultural policy has been a powerful force molding the type of farming operations we have in the United States (Donham and Thu 1993; 1995). Furthermore, agricultural policy may be implicated as a factor in a variety of agricultural health and environmental problems. For example, the major thrusts of agricultural policy for the past several decades have been toward increasing size, intensity, and specialization of farming with decreasing human resources as Ikerd describes in his contribution to this book. One specific example of such a policy is the five-year depreciation schedule on livestock buildings in the 1970s and 1980s. This allowed a producer to build a $500,000 building and deduct $100,000 per year off his taxable income. This and other policies have fostered the expansion of confinement animal production systems and associated negative agrarian health conditions for farmers, farm workers, and their families. This is but one example of farmers yielding to forces causing them to expand and specialize production in order to maintain their incomes, which is in turn linked to the adoption of new production methods based on research from land grant universities. The human and environmental health consequences of many of these technologies are clear. Over the last four years, there have been at least nineteen deaths in Iowa and surrounding states directly linked to confined swine production facilities (Donham et al. 1982). The following case exemplifies these tragedies.

A farmer in eastern Iowa accompanied by his two sons was attempting to pump out a manure storage pit. The pump stopped working, and the twenty-three-year-old son climbed into the pit to investigate. He fell off the pump into the bottom of the pit. The twenty-five-year-old son climbed into the pit to rescue his brother and also fell unconscious. The father met the same fate when he entered the pit in an attempted rescue before the hired man called the local volunteer fire department. With the aid of self-contained breathing apparatuses, firemen were able to retrieve all three men, all pronounced dead on arrival at the local hospital emergency room. High levels of hydrogen sulfide released from the agitated liquid manure was the cause of these deaths.

Less visible because the stories do not make the headlines are the slowly developing chronic illnesses such as chronic bronchitis and occupational asthma that deprive people of their health and quality of life over time. As many as 30 percent of people working in swine confinement buildings suffer one or more chronic respiratory illnesses (Donham 1990).

Advisory panels such as the Science and Education Administration and the National Agricultural Research and Extension Users Advisory Board were formed in the 1970s in an attempt to shift agricultural science policy away from a focus on bolstering productivity and toward a consideration of other factors, such as the environment and meeting basic human needs. However, these efforts generally have failed, and there has been a continuation of earlier agricultural policies focusing on increasing production (Britan 1987).

Current policies are not fostering a healthy agricultural occupation. Sustainable agriculture must include the health of the people living on the land, as well as the maintenance of natural resources. Sustainable agriculture may be defined as "leaving the land better than when it was put into our custody," (Strange 1984a) and "nourishing a renewable pool of human land stewards who earn a healthy living by farming well" (Strange 1984b). These land stewards and their families now work in a system associated with unacceptable rates of injury, accidental death, illness, and personal loss (Brown 1989). The human resources in agriculture are being stressed by hazards inherent in our production systems.

THE HUMAN HEALTH COSTS OF INTENSIVE CONFINEMENT PRODUCTION

Since the 1950s, agricultural enterprises in most Western countries have become larger, more intensive, more specialized, more capital intensive, and less labor intensive (Donham and Thu 1993). Confinement livestock production is an example of one such system developed to raise large numbers of animals in a relatively small, confined space with minimal labor (Strange 1984b). Compared to conventional livestock housing, the typical confinement building is more enclosed and tightly constructed. A higher density of animals is housed in these buildings, usually for twenty-four hours a day, starting from birth until they are market ready. Confinement housing systems must ventilate and heat the building, feed and water the animals, and hold wastes.

Manure is handled by one of two systems: it either drops through slatted floors into a pit beneath the facility until it is pumped out to be spread on fields, or it is removed through one of several mechanisms to a storage pit or lagoon outside the building. Outside storage is typical of most newer systems, but a large number of buildings remain in operation with storage pits directly below the building.

Intensive confinement systems first developed in the 1950s in U.S. poultry production. Confined swine production began in Europe in the early 1960s and in North America in the late 1960s and early 1970s. Intensive swine and poultry production facilities have now started to appear in Mexico, South America, Southeast Asia, and the Pacific Rim, including Taiwan, the Philippines, Thailand, and Vietnam. Today, sheep, beef cattle, dairy cattle, and veal calves may be housed in confinement buildings, though less commonly than swine and poultry. In contrast to other livestock, swine and poultry confinement buildings are typically fully enclosed, resulting in higher concentrations of interior environmental contaminants.

In the late 1970s in the United States, an estimated 700,000 persons worked in livestock and poultry confinement operations (Donham and Mutel 1982), representing about 50 percent of total swine production operations. This number includes owner-operators, spouses, children, employees, veterinarians, and a variety of service workers. Workers in these facilities typically prepare feed, feed animals, clean the buildings, sort and move animals, perform vaccinations or other treatments, and work at various other management and maintenance procedures.

Daily Work in a Swine Confinement Operation

Although the work day in a swine confinement operation depends on a myriad of factors, a typical day for a medium-sized farrow-to-finish family operation (250 sows) resembles the following:

5:00 A.M. John the farmer rises, dresses, eats breakfast, and goes down to the swine facility.

6:00 A.M. John observes a sow having difficulty giving birth. He calls his house and asks his wife, Jane, to help out. She helps deliver ten live pigs while he injects the sow with antibiotics and other medications. Jane returns to the house, finishes breakfast for the kids, and gets them off to school on the bus.

7:00 A.M. For twenty minutes, John hand feeds eighty hungry sows while they make ear splitting racket (over one hundred decibels; ninety decibels is the OSHA-recommended limit). The feed dust gets into his eyes, nose, and respiratory tract, producing a cough and sore throat.

8:00 A.M. Jane comes back to the farrowing barn to help him "process" the five hundred baby pigs born the day before. For two hours, she clips their canine teeth, castrates male pigs, docks tails, gives iron injections, and notches ears to identify future brood stock. She cuts her hand with a scalpel while castrating, and John accidentally inoculates himself with the iron injection. They wipe their wounds off the best they can and continue working.

10:00 A.M. John and Jane go into the nursery, a second-stage growth area where they place weaned baby pigs. Jane helps move four hundred forty-pound pigs to the grower unit—the third and final growth stage area. Jane brings in the high pressure sprayer and washes down the four rooms from which the pigs were removed. She gets dripping wet and inhales dust particles forced into the air by the spray blast. Meanwhile, John feels tightness in his chest after they move the pigs.

12:00 P.M. John and Jane grab a quick sandwich and soda in the office area. They head to the gestation room and face eighty sows in estrus and decide they are ready to breed. John moves fifty sows to the breeding stall and artificially inseminates each with frozen semen. He then moves thirty sows not in estrus to the boars' pen and observes that twenty of them will "stand" (be naturally serviced by the boars). The sows are hungry and very noisy at one hundred decibels. John begins to cough from the extra dust stirred up. A sow steps on Jane, and a very active boar knocks her down.

3:00 P.M. John and Jane go to the finishing barn, the final growth area where hogs are readied for market. They repair ten feeders and eight waterers. They sort one hundred pigs weighing approximately 240 pounds each from ten pens of twenty animals each. The pigs are put into a holding pen to be ready for transportation to market the next morning. John develops a "sinus" headache from the dust the animals kick up.

5:00 P.M. John and Jane return to the grower room where they vaccinate eighty fifty-pound pigs. John accidentally inoculates Jane with a vaccine when a pig she is holding suddenly tries to jerk itself free.

6:00 P.M. They go back to the nursery, pull all the plugs which drain the manure from the shallow pits, run the scrapers under the grower to push the manure away, and flush all subfloors and shallow pits with water.

7:00 P.M. It is time for dinner.

8:00 P.M. Check sows due to have baby pigs (farrowing).

9:00 P.M. End of the work day, return to the house for the night.

10:00 P.M. Jane develops flulike symptoms with fever, muscle aches, and headache. She recognizes this familiar condition (known as organic dust toxic syndrome) as something that occurs every time she power washes. She knows she will suffer for a day or so before it goes away.

This typical work day shows a number of health exposures in these buildings both for humans and for animals. However, there are significant animal health problems that lead to economic concerns as well (Donham 1990; Donham et al. 1982). Reports in 1977 first indicated the presence of health hazards to persons working in confined swine production environments (Donham 1990; Donham et al. 1982). Bioaerosols, fine airborne dust containing bacteria and bacterial products, are a major problem. In livestock facilities these are a complex mixture (see Table 4.1) that originate primarily from the animals, dried feces and feed (Donham et al. 1985). The decomposition of animal urine and feces produces gases, most notably ammonia and hydrogen sulfide, that affect the health of workers (Donham 1990).

Bioaerosol particles contain approximately 25 percent protein. One-third of these breathed particles are so small they are not filtered out in the nose and

Table 4.1
Potentially Hazardous Bioaerosol Agents in Swine Confinement Buildings

Microbes and their metabolites
- Endotoxin
- 1→3 β (D)Glucan
- Mycotoxins (e.g. aflatoxin, fumonasin)
- Microbial proteases

Infectious agents
Feed particles: grain dust, antibiotics, and growth promotants
Tannins
Dried livestock or poultry proteins (urine, dander, serum)
Swine feces
Grain mites, insect parts
Mineral ash
Ammonia absorbed to particles
Pollen
Mold (spores, sporangia, hyphae)
Bacteria

throat and enter the airways; the smallest particles reach the lungs. Particles from fecal material include high concentrations of gut-flora bacteria and gut-lining cells from the pig feces. Since these particles are quite small relative to other bioaerosol components, they constitute the major burden to small airways and the lungs. The larger particles are mainly from feed grain and are a problem in the upper airways of the nose, sinus cavities, and throat. Also present are animal dander, broken bits of hair, bacteria, inflammatory substances from bacteria known as bacterial endotoxins and glucans, pollen grains, insect parts, and fungal spores. Bioaerosol particles absorb ammonia and possibly other toxic or irritating gases, compounding the potential hazards of inhaled particles (Donham et al. 1985). Toxic gases are generated in the manure pit, and if located under the buildings, rise into the air of the confinement building. Of the more than 40 gases generated in decomposing manure, hydrogen sulfide, ammonia, carbon dioxide, methane, and carbon monoxide are potentially hazardous gases (Donham, Carson, and Adrian 1982; Donham et al. 1982). Ammonia is also released by bacterial action in urine and feces on confinement house floors. Carbon monoxide and carbon dioxide are generally not produced in hazardous concentrations from manure pits, but they are produced in much higher concentrations by fuel-burning heating systems used in cold weather.

Bioaerosol and gas concentrations increase in winter when facilities are tightly closed and ventilation rates are reduced to conserve heat. Typical ventilation systems in swine buildings are designed to control heat and humidity but do not necessarily reduce bioaerosol or gas levels adequately to insure a healthy environment for humans or animals. During the cold seasons, should the ventilation systems fail for several hours, carbon dioxide from animal breathing and carbon monoxide from heaters and manure pits can rise to deadly levels.

Hydrogen sulfide gas is most hazardous when the manure pits are located directly beneath the production buildings. However, gases from outside pits backflowing into a building, or any confined space where this gas can be trapped, can produce an acutely toxic environment. Hydrogen sulfide gas poses a potentially lethal hazard when the liquid manure is agitated, which operators commonly do to suspend the solids so they can pump the pits empty. During agitation, hydrogen sulfide gas can be rapidly released, increasing from usual levels of less than five parts per million (ppm) to lethal levels of over five hundred ppm within seconds. Numerous cases have been reported of workers and animals dying or becoming seriously ill when hydrogen sulfide levels increased from agitated pits.

Sixteen studies were reviewed by Rylander and others (1989) involving approximately three thousand swine confinement workers over the preceding fifteen years. Others have substantiated these findings (Cormier et al. 1991; Zuskin et al. 1991). Figure 4.1 compares the major chronic respiratory symptoms among swine confinement workers in the United States, the Netherlands,

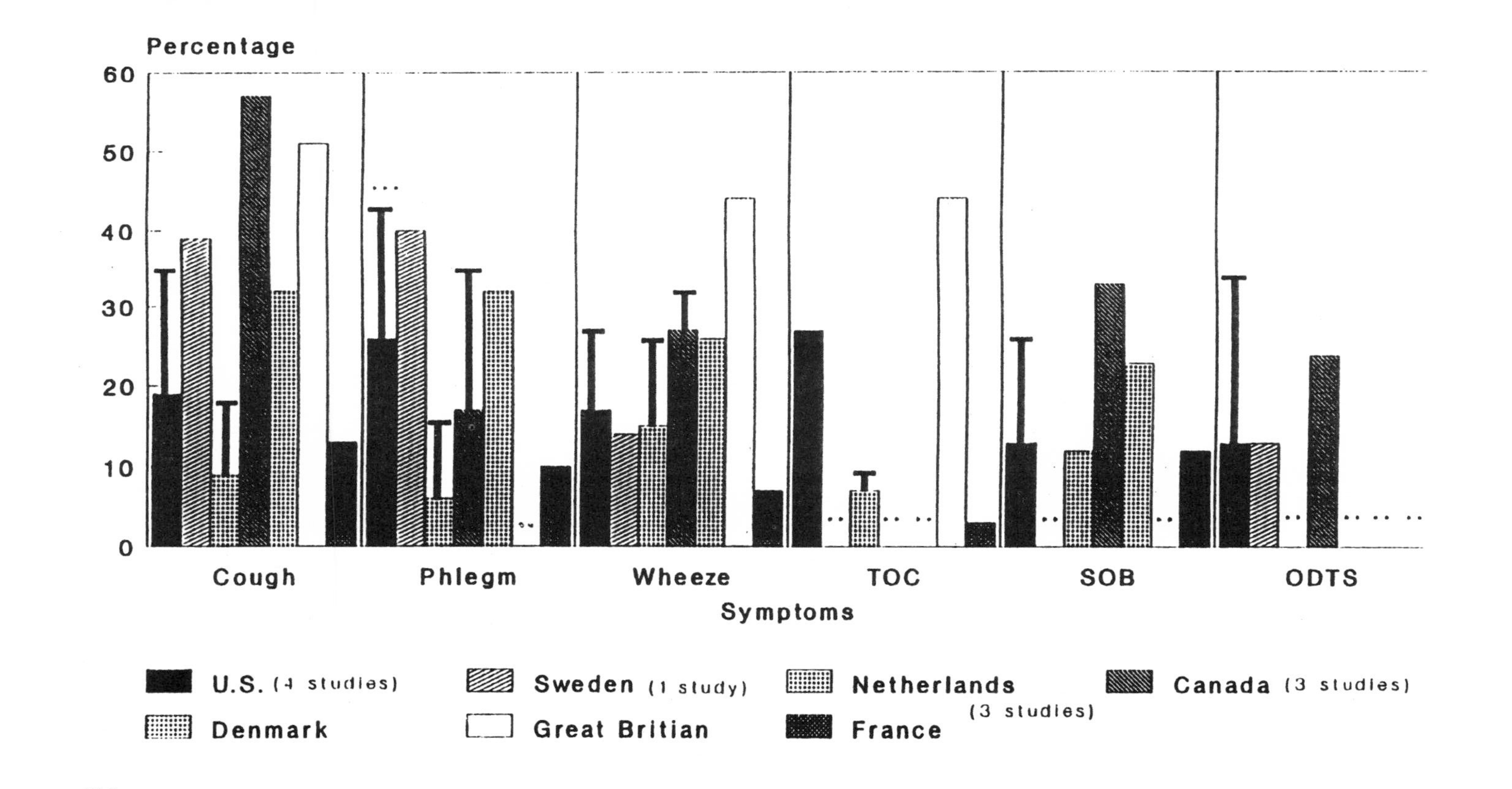

Figure 4.1

Chronic Respiratory Symptoms among Swine Confinement Workers in Different Countries*

Sweden, and Canada. Symptoms occurred in swine confinement workers two to four times more often than in comparison populations. Cough and phlegm (symptoms of bronchitis) were the two most common symptoms, occurring in 12 to 55 percent of the population. The extent of bronchitis-like symptoms was similar in the United States, Sweden, and Canada. Bronchitis was reported about 50 percent less frequently in the Netherlands compared to other countries. However, the prevalence of wheezing and chest tightness (symptoms of asthma) in the Netherlands was similar to that in the other three countries.

Organic dust toxic syndrome (ODTS) is an acute influenzalike illness that follows four to six hours of intense exposure to agricultural dusts. It is characterized by malaise, weakness, muscle aches, fever, chest tightness, and headache that may last from twelve to seventy-two hours. Rates of ODTS in the United States, Canada, and Sweden are similar, affecting about 20 percent of swine confinement workers. Workers commonly report chronic fatigue accompanied by episodes of dizziness and fainting. Shortness of breath might also be a part of this syndrome. Auger, Gourdeau, and Miller (1994) described a similar clinical pattern in people exposed to airborne molds. This evidence suggests that swine producers are experiencing a chronic toxic syndrome, possibly a chronic form of ODTS.

Acute symptoms, defined as those which workers themselves directly associate with their working environment, have also been studied (Brouwer et al. 1986). Swine confinement workers in the United States, Sweden and Canada have similar rates of acute bronchitis—approximately 55 percent—about twice the rate of chronic bronchitis. Approximately 35 percent of confinement workers experience acute symptoms of wheezing and chest tightness—symptoms of asthma.

Respiratory testing (pulmonary function tests) of swine confinement workers suggests that some experience obstructive lung disease. This finding is substantiated by U.S. studies indicating that air becomes entrapped in the lungs because chronic narrowing of the airway obstructs exhaling (Schwartz et al. 1990). This condition is commonly seen in long-term, heavy cigarette smokers. Research in Sweden and the United States examined changes in respiratory tests and found statistically significant declines indicating the presence of occupational asthma (Donham 1990; Haglind and Rylander 1987). Twenty percent of swine farmers in the Netherlands and 14 percent of swine workers in Canada had clinically significant pulmonary function declines (see table 4.2).

Workers who are in confinement buildings for more than two hours per day for six or more years are at higher risk for severe respiratory problems. Most small confinement facility owner/operators may work only a few hours per day in these facilities because they are part of a diversified operation. However, many large facility managers and hired hands work in these facilities eight hours or more a day five to seven days per week. As specialization and

Table 4.2
Exposure Thresholds in Swine Buildings

	Human Health	*Swine Health*
Total dust mg/m^3	2.4	3.7
Respirable dust mg/m^3	0.23	0.23
Endotoxin g/m^3	0.08	0.15
Carbon dioxide (ppm)	1,540	1,540
Ammonia (ppm)	7.0	11.0
Total microbes cfu/m^3	4.3×10^5	4.3×10^5

intensification of swine production continues, an increasing number of workers can be expected to spend even more time in these operations with long-term health consequences.

CONCLUSION

Agricultural policies have fostered production systems that have inherent worker and environmental health hazards. A sustainable agricultural policy must include the health and welfare of workers and farmers as an integral part of the system. The way to improve the health of farmers and farm workers is to strengthen policy forces that lead toward sustainability in agriculture. As Marty Strange says (1984a), "there must be a mechanism within the system that provides assessment and disclosure of the health impact of policy and technology." It may be helpful for a special group outside the "industry" and its land grant university supporters to study the system and develop long-range plans beyond the scope of short-term economic, political, and trading considerations. Such a plan would, of necessity, include consideration of agricultural health and welfare of farmers and their neighbors, while promoting the sustainability of agriculture and rural communities. Achieving such a goal would require the efforts of all those affected and would have to take place in an atmosphere free from the pressures of special interest groups, including production-oriented farm groups and environmentalists.

REFERENCES

Auger, Pierre L., Pierre Gourdeau, and J. David Miller. 1994. Clinical experience with patients suffering from a chronic fatigue-like syndrome and repeated upper respiratory infections in relation to airborne molds. *American Journal of Industrial Medicine* 25: 41–42.

Britan, Gerald M. "Politics of Agricultural Science." 1987. In *Farmwork and Fieldwork: American Agriculture in Anthropological Perspective,* ed. M. Chibnik. Ithaca, N.Y: Cornell University Press.

Brouwer, R., et al. 1986. Respiratory symptoms, lung function and IgG_4 levels against pig antigens in a sample of Dutch pig farmers. *American Journal of Industrial Medicine* 10: 283–285.

Brown, B. 1989. *Lone Tree: A true story of murder in America's heartland.* New York: Crown Publishing, Inc.

Cormier, Y., et al. 1989. Relationships of air quality and productivity in intensive swine housing. *Agri-Practice* 10: 15–26.

———. 1991. Respiratory health of workers exposed to swine confinement buildings and dairy barns. *Scandinavian Journal of Work and Environmental Health* 17: 269–275.

Donham, Kelley. 1990. Health effects from work in swine confinement buildings. *American Journal of Industrial Medicine* 4: 17–26.

Donham, Kelley, T. L. Carson, and B. R. Adrian. 1982. Carboxyhemoglobin values in swine relative to carbon monoxide exposure: Guidelines to monitor for animal and human health hazards in swine buildings. *American Journal of Veterinary Research* 5: 813–816.

Donham, Kelley, and Connie Mutel. 1982. Agricultural medicine: The missing component of the rural health movement. *Journal of Family Practice* 14: 511–520.

Donham, Kelley, and Kendall M. Thu. 1993. Relationships of agricultural and economic policy to the health of farm families, livestock, and the environment. *Journal of the American Veterinary Medical Association* 202(7): 1084–1091.

———. 1995. "Agricultural Medicine and Environmental Health: The Missing Component of the Sustainable Agricultural Movement." In *Human Sustainability in Agriculture.* ed. H. H. McDuffie, et al. Boca Raton, Fla.: Lewis Publishers.

Donham, Kelley, L. W. Knapp, Russell Monson, and Kim Gustafson. 1982. Acute toxic exposures to gases from liquid manure. *Journal of Occupational Medicine* 24: 142–145.

———. 1985. Characterization of dusts collected from swine confinement buildings. *American Industrial Hygiene Association Journal* 46: 658–661.

Haglind, P., and R. Rylander. 1987. Occupational exposure and lung function measurements among workers in swine confinement buildings. *Journal of Occupational Medicine* 29: 904–907.

Merchant, James, et al. 1989. *Agriculture at Risk: A Report to the Nation.* Kansas City, Mo.: The National Coalition for Agricultural Safety and Health, the National Rural Health Association.

National Safety Council. 1994. *Accident facts.* Chicago: NSC.

Rylander, R., et al. 1989. Effects of exposure to dust in swine confinement buildings—A working group report. *Scandinavian Journal of Work Environmental Health* 15: 309–312.

Schwartz, David A., et al. 1990. Are workshift changes in lung function predictive of underlying lung disease? Paper presented at the World Conference on Lung Health, American Lung Association and American Thoracic Society.

Strange, Marty, ed. 1984a. *It's not all sunshine and fresh air: Chronic health effects of modern farming practices.* Walthill, Nebr.: Center for Rural Affairs.

———. "The Economic Structure of a Sustainable Agriculture." In *Meeting the Expectations of the Land,* ed. W. Jackson, W. Berry, and B. Colman. San Francisco: North Point Press, 1984b.

Zuskin, E., et al. 1991. Immunological and respiratory findings in swine farmers. *Environmental Research* 56: 120–30.

Chapter 5

Mood Changes Experienced by Persons Living Near Commercial Swine Operations

Susan S. Schiffman, Elizabeth A. Sattely-Miller, Mark S. Suggs, and Brevick G. Graham

INTRODUCTION

Several years ago a woman from rural North Carolina came to our laboratory in a very distraught emotional state. She stated that her depression was due to odors from a new swine operation near her home. At that time, we were studying the effect of fragrances on mood. We decided to extend our research to evaluate the effects of swine odor on mood as well. We conducted a controlled scientific study to determine if swine odor indeed altered people's moods. The results showed highly significant mood disturbances in persons living downwind from swine operations.

Odor has always been associated with livestock and poultry farming (Cox 1975; Lehmann 1973; Miner 1974; Nielsen, Voorburg, and L'Hermite 1991; 1986; Pain et al. 1991; Rains, Deprimo, and Groseclose 1973). Today, however, large confined livestock operations are producing odors that are a major environmental management challenge and are raising health concerns among neighbors (Barth and Melvin 1984; Casey and Hobbs 1994; Kilman 1994; Miller 1994; Waggoner 1994; Walsh 1994; Warburton et al. 1980; Warner, Sidhu, and Chadzynski 1990; Williams et al. 1989). Environmental odors can affect a population's physiological and psychological well-being (Rotton 1983; Shusterman 1992; Winneke and Kastka 1977).

Unpleasant odors can affect the well-being of people by "eliciting unpleasant sensations, triggering possible harmful reflexes, modifying olfactory function and other physiological reactions" (Miner 1980). Miner also points out that people can experience annoyance and depression, along with nausea, vomiting, headache, shallow breathing, coughing, sleep disturbances, and loss of appetite in response to unpleasant odors. Even low concentrations of odor-producing compounds from livestock operations are likely to produce complaints (Carney and Dodd 1989; Laing, Eddy, and Best 1994).

Neutra and Colleagues (1991) found that residents living near a hazardous waste site who complained of odors had more symptoms than those who did not complain, regardless of distance from the site. In a review of municipal, agricultural, and industrial settings (e.g. Ames and Stratton 1991; Goldsmith 1973; Jonsson and Sanders 1975; Satin et al. 1986; Satin et al. 1987; Scarborough et al. 1989), Shusterman (1992) also reports a strong relationship between reported symptoms and exposure to odor.

Odors associated with swine operations come from a mixture of urine, fresh and decomposing feces, and spilled feed. These odors come from air ventilated from confinement buildings and waste storage and handling systems—including manure lagoons—and field fertilization of pig manure (Bundy 1991). Anaerobic (meaning absence of free oxygen) microbial decomposition of feces appears to be the source of more objectionable smells (Ritter 1989). Compounds identified in livestock manure include sulfides, disulfides, volatile organic acids, alcohols, aldehydes, amines, fixed gases, nitrogen heterocycles, mercaptans, carbonyls, and esters (Du Toit 1987; Miner 1980; 1975; Miner, Kelly, and Anderson 1975; Skarp 1975). It is likely that it is the mixture of compounds, rather than only one component, that contributes to the mood changes measured here.

Scientists have developed models that measure dispersion of odors in the air to predict the peak and mean concentrations of odors and environmental air pollutants at various distances from the source (Cha, Li, and Brown 1994; Gassman 1992; Janni 1982; Nordstedt and Taiganides 1971), and patterns of complaints at varying distances have been studied (Clarenburg 1987). Concentrations of odorous compounds that flow in a plume are not significantly reduced over 750 to 1500 feet downwind of the source (Gassman 1992). Numerous techniques for reducing odor from livestock operations have been evaluated, with disappointing results (Warburton et al. 1980). To date, aerobic treatment is the most effective method for deodorizing pig waste slurry (Al-Kannani et al. 1992; Beaudet et al. 1990, Bourque et al. 1987; Lau et al. 1992; Sneath 1988; Sneath et al. 1992; Sneath and Williams 1990; Williams et al. 1989).

Our study uses a well-standardized scale to quantify objectively the moods of people exposed to odors near large-scale hog operations. The Profile of Mood States questionnaire (McNair et al. 1992; McNair and Lorr 1964) was employed to assess mood among individuals living near swine operations and in control participants. The scale has been used in a number of situations, including studies evaluating the effect of pleasant odors on mood (Schiffman et al. 1994a; Schiffman et al. 1994b). Investigating mood in persons exposed to odors is an important health issue; negative mood can affect immunity (Calabrese et al. 1987; O'Leary 1990; Stone et al. 1987; Weisse 1992) which can influence susceptibility to disease.

METHOD

Forty-four residents living near hog operations and forty-four control participants not living near hog operations in rural North Carolina were studied. Participants in the control and experimental groups were matched by gender, age, race, and years of education. Twenty-six participants in each group were female, and eighteen in each were male. In both groups, the majority of participants were employed as skilled laborers. In addition, control and experimental groups were matched for number of chronic illnesses; fourteen participants in each group suffered allergies. Average time of residence near hog operations was 5.3 years, with a range of 8 months to 27 years.

All participants filled out Profile of Mood States (POMS) questionnaires, as well as consent forms and a general information questionnaire on demographic, dietary, and medical history. Mood ratings were obtained from the POMS, chosen because of its demonstrated sensitivity to mood shifts (McNair et al. 1992; McNair and Lorr 1964). POMS covers sixty-five descriptive mood categories. Most are grouped by six factors: tension-anxiety, depression-dejection, anger-hostility, vigor-activity, fatigue-inertia, and confusion-bewilderment. The descriptive adjectives for each of the six factors are given in table 5.1. The scale for rating each feeling ranges from 0 (not at all) to 4 (extremely). Ratings for each factor are added to get factor totals; then factor totals are added to compute a Total Mood Disturbance score (TMD).

Experimental participants living in the vicinity of swine operations were asked to fill out one POMS questionnaire on each of four days when hog odors could be smelled. The days did not have to be consecutive, and there was no time limit for completing the four POMS questionnaires. Participants not living near hog facilities were asked to fill out one POMS on each of two days. Both groups were asked to complete the POMS questionnaires based on how they were feeling recently, including at the time they responded to the questionnaire.

Table 5.1
Adjectives that describe each factor of the POMS scale

Tension-Anxiety	*Depression-Dejection*	*Anger-Hostility*
Tense	Unhappy	Angry
Shaky	Sorry for things done	Peeved
On edge	Sad	Grouchy
Panicky	Blue	Spiteful
Relaxed (-)	Hopeless	Annoyed
Uneasy	Unworthy	Resentful
Restless	Discouraged	Bitter
Nervous	Lonely	Ready to fight
Anxious	Miserable	Rebellious
	Gloomy	Deceived
	Desperate	Furious
	Helpless	Bad-tempered
	Worthless	
	Terrified	
	Guilty	
Vigor-Activity	*Fatigue-Inertia*	*Confusion-Bewilderment*
Lively	Worn-out	Confused
Active	Listless	Unable to concentrate
Energetic	Fatigued	Muddled
Cheerful	Exhausted	Bewildered
Alert	Sluggish	Efficient (-)
Full of pep	Weary	Forgetful
Carefree	Bushed	Uncertain about things
Vigorous		

RESULTS

Figure 5.1 and table 5.2 outline the findings. Figure 5.1 illustrates the response averages for the experimental versus the control group for all POMS factors and the (TMD). Table 5.2 summarizes results of the statistical analysis performed to determine whether the differences between groups were significant or the result of random variation. For all of the POMS factors and the TMD combined, there was a statistically significant difference ($p < 0.0001$) between the experimental group (people living near swine operations) and the control group (those not living near swine operations). The experimental group had significantly worse mood scores than the control group for every POMS factor and the TMD. It is statistically improbable that these differences can be attributed to factors other than those associated with the swine operations. In addition, men living near swine operations had significantly worse

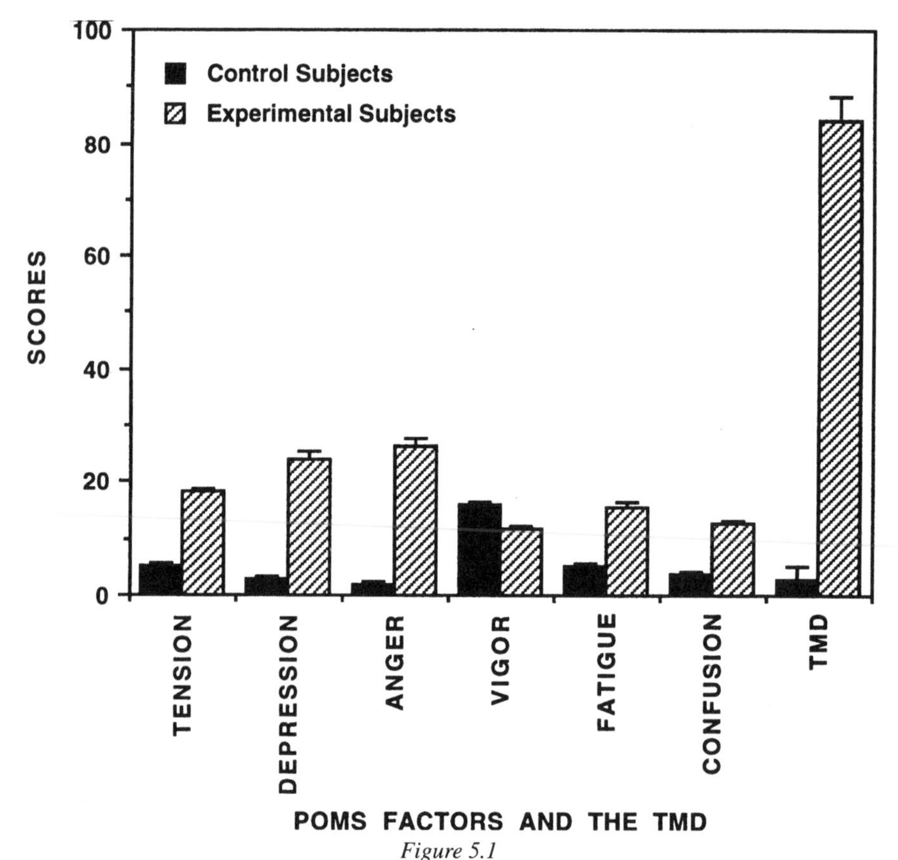

Figure 5.1

Mean POMS scores of each factor and the total mood disturbance score (TMD) for experimental subjects who live near swine operations and control subjects

scores for anger ($p < .01$) than women. Men also had higher confusion scores ($p < .005$) than women living near swine operations, as well as higher scores than both men and women in the control group not living near swine operations. Women living near swine operations also had higher confusion ratings than either men or women living away from these operations. Scores for men and women living away from swine operations did not differ significantly from each other.

Our study found that persons living near swine operations who smelled odors were significantly more angry, depressed, tense, fatigued, confused, and less vigorous than control participants not living near swine operations. Those exposed to odors also had more total mood disturbance than controls. These

Table 5.2
Results

Effect	*Tension*	*Depression*	*Anger*	*Vigor*	*Fatigue*	*Confusion*	*Total Mood Disturbance Score*
Group	k	k	k	k	k	k	k
Gender			k				
Group × Gender						k	
Subject (Group, Gender)	k	k	k	k	k	k	k

k Significant at $\mu = 0.05$ level

findings are consistent with previous studies in which pleasant and unpleasant odors were found to affect mood (Baron 1990; Ehrlichman and Bastone 1992; Rotton 1983; Schiffman et al. 1994a 1994b; Shusterman 1992; Winneke and Kastka 1977). In different settings, odors can affect cognitive performance (Lorig 1992; Ludvigson and Rottman 1989) and physiological responses such as heart rate and electroencephalagraphic patterns (Lorig 1989; Lorig et al. 1991; Lorig and Roberts 1990; Manley 1993).

COMPARISON WITH OTHER STUDIES

Two additional studies (Schiffman 1994–1995; unpublished) provide further support for the finding that odors from hog operations can impair mood. Since the POMS scale contains more negative feelings than positive ones, two different types of mood scales were used in the new studies. In study 1, a mood scale consisting of fifty adjectives containing equal numbers of positive and negative feelings was administered to persons living near hog operations (experimental participants) as well as control participants not living near hog operations. In study 2, a simple mood scale labeled "very good" at one end and "very bad" at the other was used; the midpoint was labeled "neither good nor bad." In each of these two studies, mood judgments were made daily over a two-week period. For study 1, participants were asked to complete mood questionnaires at 9:30 P.M. In study 2, participants were asked to mark the mood scales at 8:00 A.M. These times of the day were selected for recording mood because odors from hog operations tend to be strongest in the early morning and after sundown. Participants in both studies also completed a questionnaire asking for weather information and to determine whether they smelled odors that day. Participants were not told the purpose of either study other than it was an evaluation of rural mental health.

The results of these two additional studies support the findings of the present report using the POMS scale. On days when experimental participants experienced odors, their moods were substantially different from the moods of control participants. In fact, experimental participants almost never recorded a positive feeling on days when they experienced hog odors. On the day following strong odors, moods were still suppressed to varying degrees. These data suggest that hog odors not only may produce negative moods, but also may prevent positive ones. In addition, negative moods resulting from hog odors on one day can persist to the next. Laboratory studies also have shown that negative moods from unpleasant odors can persist beyond the time of experience (Ehrlichman and Bastone 1992).

POSSIBLE CAUSES OF ALTERED MOOD

Factors that may play a role in the altered mood of swine operation neighbors exposed to odors include: (a) the unpleasant sensory quality of the odor; (b) uncontrollable timing of the exposures; (c) learned aversions to the smell; (d) possible neural stimulation of immune responses through direct neural connections between odor centers in the brain and lymphoid tissue; (e) direct physical effects from nasal and respiratory irritants in the odor plume; (f) possible sensory disorders; and (g) unpleasant thoughts associated with smells.

Most persons rate smells from hog farm operations as unpleasant when they are present at moderate to high intensity. Neighbors experience odors outdoors and inside their homes via open windows and air conditioning systems. The smell can permeate clothing, curtains, and building materials, which then release the odor over time after the odor-bearing plume from the swine operation has passed. The intermittent nature of the odors also may be a factor. Studies of noise show that intermittent sounds produce more arousal and are more likely to negatively affect performance than constant noise (Cohen and Weinstein 1981). Lack of control over the timing of unwanted stimuli contributes to the negative impact. Humans and animals respond differently to irregular versus predictable noise (De Boer et al. 1989).

The psychological and physical effects of odors may be a learned response; conditioned aversions to smells are well-documented in the scientific literature (Dyck et al. 1990; Goodwin et al. 1992; Hunt et al. 1993; Meachum and Bernstein 1992; Murua and Molina 1990; von Kluge and Brush 1992). Environmental odors associated with irritants or other toxic chemicals may lead to negative conditioning (Shusterman 1992). The conditioned alterations in immune responses based on smell and taste stimuli are strong evidence for links between chemosensory centers in the brain and the immune system (Ader and Cohen 1991). Suppression and enhancement of the immune system can be conditioned responses to chemosensory stimuli (Ader and Cohen 1991; Dyck et al. 1990; Hiramoto et al. 1991; Hiramoto et al. 1993; Solvason et al. 1991; Solvason et al. 1992).

Unpleasant odors may affect physical health through the direct anatomical connections between the olfactory and the immune systems. The brain's integrated circuitry between the limbic cortex, limbic forebrain, hypothalamus, brain stem, and brain structures involving smell and emotion can influence immune responses (Brodal 1992; Brooks et al. 1982; Cross et al. 1982; Cross et al. 1980; Fukuda et al. 1987; Guevara-Guzman et al. 1991; Kadlecova et al. 1987; Karadi et al. 1989; Katayama et al. 1987; Nance et al. 1987; Ono et al. 1985; Oomura 1980; Oomura 1987; Oomura et al. 1970; Roszman et al. 1982; Scott and Pfaffmann 1967; Sutin et al. 1975; Takagi 1986; Tanabe et al. 1975; Tazawa et al. 1987; Tyrey and Nalbandov 1972). Brain–immune system links work both ways (Solomon 1987), making it possible for immune responses to affect odor centers in the brain (Besedovsky et al. 1977; Saphier et al. 1987).

Specific molecules in the odorous plume from hog operations may have direct physical effects on the body, including nasal and respiratory irritation (Bundy 1991; Cometto-Muniz and Cain 1991; Donham 1990; Miner 1980; Shusterman 1992). Nasal irritation can elevate adrenalin (Allison and Powis 1976), which can contribute to anger and tension. The volatile organic compounds responsible for odors can be inhaled and transferred into blood and body fat (Ames and Stratton 1991; Ashley et al. 1994; Lam et al. 1990; Mage 1991; Wilcosky 1993). These compounds may be released over time, so that the exposed person continues to smell the odor after the plume carrying it has passed. Volatile organic compounds (VOCs) are known to be eliminated in the breath after exposure (Raymer et al. 1991; Wallace et al. 1991), and methods for measuring VOCs in breath have been documented (Phillips and Greenberg 1992; Raymer et al. 1991; Thomas et al. 1991). Theoretically, it would be possible for some compounds in the odor plume to reach the brain by way of olfactory neurons since a range of airborne agents has been found to do so via the nasal pathway (Delorenzo 1970; Esiri and Tomlinson 1984; Jackson et al. 1979; Monath et al. 1983; Roberts 1986; Shipley 1985). The endotoxin produced by bacteria found in the swine operation air also may be present in the odorous plume. Individuals with abnormal smell functioning caused by factors not related to swine operations, including medical conditions, medications, or pesticide exposures (Schiffman 1983), may find the smell even more objectionable.

Finally, smells can alter moods. The odor of a hog farm may be taboo for some, while for others the complex overlays of fear for their property values, concern about the environment, and personal discomfort in the daily pursuit of enjoying their life and property affect outlook. Smells from livestock operations may be considered inappropriate in particular environments. Odor complaints occur most frequently for new, large, or recently expanded operations near existing residences or shopping areas (Miner 1980; Sweeten and Miner 1993). Swine farm odor complaints may be part of an increased awareness of other environmental agents, such as tobacco smoke.

CONCLUSION

Smells from large hog operations have a significant negative impact on the mood of nearby residents. To minimize the impact on the lives of swine operation neighbors, ways must be found to reduce the concentrations of the volatile organic compounds responsible for odor. This may be in the form of legislated standards and technology to treat odorous compounds.

The Clean Air Act does not regulate odors, which are generally regarded as nontoxic (Bundy 1991). State-level legislation for controlling odors is imprecise or entirely lacking in many states. For example, the North Carolina Administrative Code Title 15A-02D.0522(c) specifies: "A person shall not cause, allow, or permit any plant to be operated without employing suitable measures for the control of odorous emissions including wet scrubbers, incinerators, or such other devices as approved by the Commission." This is a subjective regulation; it has no emissions standards or ambient air standards that can be measured or monitored. On the face of it, the regulation says a plant can emit offensive odors as long as it has suitable control devices. Moreover, it does not specify what a "plant" is. In contrast, Connecticut sets specific standards for odor emissions (Campbell et al. 1973). Likewise, in the Netherlands, regulations for livestock production controls are based on accurate records of manure production and bookkeeping and violations are considered a criminal offense (Brussard and Grossman 1990).

Regulations are needed in all fifty U.S. states because animal wastes have high levels of volatile organic compounds that can produce strong odors. The 1987 estimate for animal manure production in the United States was 1.5 billion tons per year, enough to apply almost one ton to each of the 1.9 billion acres in the continental United States (Brussard and Grossman 1990). This is a measure of the industry's challenge.

Nuisance laws have been the general avenue of remedy used by persons exposed to high levels of odors from agricultural operations. Nuisance laws, however, usually consider: (a) who was there first; (b) the neighborhood; (c) the reasonableness of the land use; and (d) kind and degree of the interference with use and enjoyment of the complainants' property (Hamilton and

Bolte 1988). In the United States, as in many other countries, agriculture has special recognition and often protection from nuisance law suits in the form of state right-to-farm statutes (Artis and Silvester 1986; Hamilton and Bolte 1988). Smells perceived as nuisances are less successful vehicles for court action than water pollution (Means 1992) and come under state law rather than federal statute, the usual venue for water pollution suits.

REFERENCES

Ader, R., and N. Cohen. 1991. "The influence of conditioning on immune responses." In *Psychoneuroimmunology,* 2d ed., ed. R. Ader, D. L. Felten, and N. Cohen, 611–646. San Diego: Academic Press.

Al-Kanani, T., E. Akochi, A. F. MacKenzie, I. Alli, and S. Barrington. 1992. Waste management: Odor control in liquid hog manure by added amendments and aeration. *Journal of Environmental Quality* 21: 704–708.

Allison, D. J., and D. A. Powis. 1976. Early and late hind-limb vascular responses to stimulation of receptors in the nose of the rabbit. *Journal of Physiology* (Lond) 262(2): 301–17.

Ames, R. G., and J. W. Stratton. 1991. Acute health effects from community exposure to *n*-propyl mercaptan from a ethoprop (Mocap)-treated potato field in Siskiyou County, California. *Archives of Environmental Health* 46: 213–217.

Artis, D., and S. Silvester. 1986 Odour nuisance: Legal controls. *Journal of Planning and Environmental Law* (Great Britain) August 565–577.

Ashley, D. L., M. A. Bonin, F. L. Cardinali, J. M. McCraw, and J. V. Wooten. 1994. Blood concentrations of volatile organic compounds in a nonoccupationally exposed US population and in groups with suspected exposure. *Clinical Chemistry* 40: 1401–1404.

Baron, R. A. 1990. Environmentally induced positive affect: Its impact on self-efficacy, task performance, negotiation, and conflict. *Journal of Applied Social Psychology* 20: 368–384.

Barth, C. L., and S. W. Melvin. 1984. "Odor." In *Agriculture and the environment,* ed. J. M. Sweeten and F. J. Humenlik, 97–106. St. Joseph, Mich.: American Society of Agricultural Engineers.

Beaudet, R., C. Gagnon, J. G. Bisaillon, and M. Ishaque. 1990. Microbiological aspects of aerobic thermophilic treatment of swine waste. *Applied Environmental Microbiology* 56: 971–976.

Besedovsky, H., E. Sorkin, D. Felix, and H. Haas. 1977. Hypothalamic changes during the immune response. *European Journal Immunology* 7: 323–325.

Bourque, D., J-G. Bisaillon, R. Beaudet, M. Sylvestre, M. Ishaque, and A. Morin. 1987. Microbiological degradation of malodorous substances of swine waste under aerobic conditions. *Applied Environmental Microbiology* 53: 137–141.

Brodal, P. 1992. *The central nervous system: Structure and function,* 370–371. Oxford: Oxford University Press.

Brooks, W. H., R. J. Cross, T. L. Roszman, and W. R. Markesbery. 1982. Neuroimmunomodulation: Neural anatomical basis for impairment and facilitation. *Annals of Neurology* 12: 56–61.

Brussard, W., and M. R. Grossman. 1990. Legislation to abate pollution from manure: The Dutch approach. *North Carolina Journal of International Law and Commercial Regulation* 15(1): 85–114.

Bundy, D. S. 1991. "Odor Issues with Wastes." In *National livestock, poultry and aquaculture waste management,* 288–292. Proceedings of the National Workshop. ASAE Publication 03–92; St. Joseph, Mich.: American Society of Agricultural Engineers.

Calabrese, J. R., M. A. Kling, and P. W. Gold. 1987. Alterations in immunocompetence during stress, bereavement, and depression: Focus on neuroendocrine regulation. *American Journal of Psychiatry* 144: 1123–1134.

Campbell, W. A., E. E. Ratliff, A. S. Boyers, and D. R. Johnston. 1973. Air pollution controls in North Carolina: A report on their legal and administrative aspects. Institute of Government, The University of North Carolina at Chapel Hill. January: 91–94.

Carney, P. G., and V. A. Dodd. 1989. The measurement of agricultural malodours. *Journal of Agricultural Engineering Research* 43: 197–209.

Casey, J. A., and C. Hobbs. "Look What the GATT Dragged In: Corporate Hogs and Toxic Waste," *New York Times,* OP-ED, 21 March 1994.

Cha, S. S., Z. Li, and K. E. Brown. 1992. A conversion scheme for the ISC model in odor modeling. Air and Waste Management Association meeting. June, Kansas City, Mo.

Clarenburg, L. A. 1987. Odour: A mathematical approach to perception and nuisance. *Developments in Toxicology and Environmental Science* 15: 75–94.

Cohen, S., and N. Weinstein. 1981. Nonauditory effects of noise on behavior and health. *Journal of Social Issues* 37: 36–70.

Cometto-Muniz, J. E., and W. S. Cain. 1991. "Influence of Airborne Contaminants on Olfaction and the Common Chemical Sense." In *Smell and Taste in Health and Disease,* ed. T. V. Getchell, R. L. Doty, L. M. Bartoshuk, and J. B. Snow, 765–785. New York: Raven Press.

Cox, J. P. 1975. *Odor control and olfaction: Handbook.* Lynden, Wash.: Pollution Sciences Company.

Cross, R. J., W. H. Brooks, T. L. Roszman, and W. R. Markesbery. 1982. Hypothalamic-immune interactions: Effect of hypophysectomy on neuroimmunomodulation. *Journal of Neurological Science* 53: 557–566.

Cross, R. J., W. R. Markesbery, W. H. Brooks, and T. L. Roszman. 1980. Hypothalamic-immune interactions: The acute effect of anterior hypothalamic lesions on the immune response. *Brain Research* 196: 79–87.

De Boer, S. F., J. Van der Gugten, and J. L. Slangen. 1989. Plasma catecholamine and corticosterone responses to predictable and unpredictable noise stress in rats. *Physiological Behavior* 45: 789–95.

DeLorenzo, A. J. D. 1970. "The Olfactory Neuron and the Blood-Brain Barrier." In *Taste and Smell in Vertebrates,* ed. G. E. W. Wolstenholme and J. Knight, 151–176. CIBA Foundation Symposium, London: Churchill.

Donham, K. J. 1990. "Health Concerns from the Air Environment in Intensive Swine Housing: Where Have We Come From and Where Are We Going?" In *Making Swine Buildings a Safer Place to Work,* 9–20. Des Moines, Iowa: National Pork Producers Council.

Du Toit, A. J. 1987. Quantification of odour problems associated with liquid and solid feedlot and poultry wastes. *Water Science Technology* 19: 31–41.

Dyck, D. G., L. Janz, T. A. Osachuk, J. Falk, J. Labinsky, and A. Greenberg. 1990. The Pavlovian conditioning of IL-1–induced glucocorticoid secretion. *Brain Behavior Immunology* 4: 93–104.

Ehrlichman, H., and L. Bastone. 1992. "The Use of Odour in the Study of Emotion." In *The Psychology and Biology of Perfume,* ed. S. van Toller and G. H. Dodd, 143–159. London: Elsevier Applied Science.

Esiri, M. M., and A. H. Tomlinson. 1984. Herpes simplex encephalitis: Immunohistological demonstration of spread of virus via olfactory and trigeminal pathways after infection of facial skin in mice. *Journal of Neurological Science* 64: 213–217.

Felten, D. L., N. Cohen, R. Ader, S. Y. Felten, S. L. Carlson, and T. L. Roszman. "Central Neural Circuits Involved in Neural-Immune Interactions." In *Psychoneuroimmunology,* 2d ed., ed. R. Ader, D. L. Felten, and N. Cohen, 3–25. San Diego: Academic Press, 1991.

Fukuda, M., T. Ono, and K. Nakamura. 1987. Functional relations among inferotemporal cortex, amygdala, and lateral hypothalamus in monkey operant feeding behavior. *Journal of Neurophysiology* 57: 1060–1077.

Gassman, P. W. 1992. Simulation of odor transport: A review. Presented at a meeting, of the American Society of Agricultural Engineers, Nashville, Tenn.

Goldsmith, J. R. 1973. "Annoyance and Health Reactions to Odor from Refineries and Other Industries in Carson, California, 1972." In *Health and Annoyance Impact of Odor Pollution,* ed. J. R. Goldsmith, 189–227. Environmental protection technology series. EPA-650/1–75–001. Washington, D.C.: Environmental Protection Agency.

Goodwin, G. A., C. J. Heyser, C. A. Moody, L. Rajachandran, V. A. Molina, H. M. Arnold, D. L. McKinzie, N. E. Spear, and L. P. Spear. 1992. A fostering study of the effects of prenatal cocaine exposure: II. Offspring behavioral measures. *Neurotoxicol. Teratol.* 14: 423–32.

Guevara-Guzman, R., D. E. Garcia-Diaz, L. P. Solano-Flores, M. J. Wayner, and D. L. Armstrong. 1991. Role of the paraventricular nucleus in the projection from the nucleus of the solitary tract to the olfactory bulb. *Brain Research Bulletin* 27: 447–50.

Hamilton, N. D. and D. Bolte. 1988. Nuisance law and livestock production in the United States: A fifty state analysis. *Journal of Agriculture Taxation and Law* 10(2): 99–136.

Heinzow, B. G. and A. McLean. 1994. Critical evaluation of current concepts in exposure assessment. *Clinical Chemistry* 40: 1368–1375.

Hiramoto, R. N., V. K. Ghanta, C. F. Rogers, and N. S. Hiramoto. 1991. Conditioning the elevation of body temperature, a host defensive reflex response. *Life Science* 49: 93–99.

Hiramoto, R. N., C. M. Hsueh, C. F. Rogers, S. Demissie, N. S. Hiramoto, S. J. Soong, and V. K. Ghanta. 1993. Conditioning of the allogeneic cytotoxic lymphocyte response. *Pharmacology and Biochemical Behavior* 44: 275–80.

Hunt, P. S., J. C. Molina, L. Rajachandran, L. P. Spear, and N. E. Spear. 1993. Chronic administration of alcohol in the developing rat: Expression of functional tolerance and alcohol olfactory aversions. *Behavioral Neural Biology* 59: 87–99.

Jackson, R. T., J. Tigges, and W. Arnold. 1979. Subarachnoid space of the CNS, nasal mucosa, and lymphatic system. *Archives of Otolaryngology* 105: 180–184.

Janni, D. A. 1982. Modeling dispersion of odorous gases from agricultural sources. *Transactions of the American Society of Agricultural Engineers* 25SE: 1721–1723.

Jonsson, E., M. Deane, and G. Sanders. 1975. Community reactions to odors from pulp mills: A pilot study in Eureka, California. *Environmental Research* 10: 249–270.

Kadlecova, O., K. Masek, J. Seifert, and P. Petrovicky. 1987. The involvement of some brain structures in the effects of immunomodulators. *Annals of the New York Academy of Sciences* 496: 394–398.

Karádi, Z., Y. Oomura, H. Nishino, and S. Aou. 1989. Olfactory coding in the monkey lateral hypothalamus: Behavioral and neurochemical properties of odor-responding neurons. *Physiological Behavior* 45: 1249–1257.

Katayama, M., S. Kobayashi, N. Kuramoto, and M. M. Yokoyama. 1987. Effects of hypothalamic lesions on lymphocyte subsets in mice. *Annals of the New York Academy of Sciences* 496: 366–376.

Kilman, S. "Power Pork: Corporations Begin to Turn Hog Business into an Assembly Line," *The Wall Street Journal,* 28 March 1994, A10.

Laing, D. G., A. Eddy, and D. J. Best. 1994. Perceptual characteristics of binary, trinary, and quaternary odor mixtures consisting of unpleasant constituents. *Physiological Behavior* 56: 81–93.

Lam, C. W., T. J. Galen, J. F. Boyd, and D. L. Pierson. 1990. Mechanism of transport and distribution of organic solvents in blood. *Toxicology and Applied Pharmacology* 104: 117–129.

Lau, A. K., K. V. Lo, P. H. Liao, and J. C. Yu. 1992. Aeration experiments for swine waste composting. *Bioresource Technology* 41: 145–152.

Lehmann, E. J. 1973. Odor pollution: A bibliography with abstracts. National Technical Information Service (NTIS)-WIN-73-033/COM-73-11463: 1–55. Environmental Protection Agency.

Lorig, T. S. 1989. Human EEG and odor response. *Progressive Neurobiology* 33: 387–398.

———."Cognitive and Non-Cognitive Effects of Odour Exposure: Electrophysiological and Behavioral Evidence." In *The Psychology and Biology of Perfume,* ed. S. Van Toller and G. H. Dodd, 161–173. London: Elsevier Applied Science, 1992.

Lorig, T. S., E. Huffman, A. DeMartino, and J. DeMarco. 1991. The effects of low concentration odors on EEG activity and behavior. *Journal of Psychophysiology* 5: 69–77.

Lorig, T. S., and M. Roberts. 1990. Odor and cognitive alteration of the contingent negative variation. *Chemical Senses* 15: 537–545.

Lorig, T. S., A. C. Sapp, J. Campbell, and W. S. Cain. 1993. Event-related potentials to odor stimuli. *Bull. Psychonom. Soc.* 31: 131–134.

Lorig, T. S., and G. E. Schwartz. 1988. Brain and odor—I: Alteration of human EEG by odor administration. *Psychobiology* 16: 281–284.

Ludvigson, H. W., and T. R. Rottman. 1989. Effects of ambient odors of lavender and cloves on cognition, memory, affect and mood. *Chemical Senses* 14: 525–536.

Mage, D. 1991. A comparison of direct and indirect methods of human exposure. *Progress in Clinical Biological Research* 372: 443–454.

Manley, C. H. 1993. Psychophysiological effect of odor. *Critical Reviews in Food, Science, and Nutrition* 33(1): 57–62.

McNair, D. M., and M. Lorr. 1964. An analysis of mood in neurotics. *Journal of Abnormal Social Psychology* 69: 620–627.

McNair, D. M., M. Lorr, and L. F. Droppleman. 1992. *Manual: Profile of mood states.* Rev. ed. San Diego, Calif. San Diego Education and Industrial Testing Service.

Meachum, C. L. and I. L. Bernstein. 1992. Behavioral conditioned responses to contextual and odor stimuli paired with LiCl administration. *Physiological Behavior* 52: 895–899.

Means, P. 1992. Eau de hog or, it smells like money to me. *Ark Lawyer* 26(2): 36–40.

Miller, M. 1994. There is something in the air. *Pork.* August: 5.

Miner, J. R. 1974. Odors from confined livestock production: A state of the art. Environmental protection technology series. EPA-660/2–74–023. Washington, D.C.: U. S. Government Printing Office.

———. 1975. "Management of Odors Associated with Livestock Production." In *Managing Livestock Wastes,* 378–380. St. Joseph, Mich.: American Society of Agricultural Engineers.

———. 1980. "Controlling Odors from Livestock Production Facilities: State-of-the-Art." In *Livestock Waste: A Renewable Resource,* 297–301. St. Joseph, Mich.: American Society of Agricultural Engineers.

Miner, J. R., M. D. Kelly, and A. W. Anderson. 1975. "Idenfication and Measurement of Volatile Compounds within a Swine Building and Measurement of Ammonia Evolution Rates from Manure-Covered Surfaces." In *Managing Livestock Wastes,* 351–353. St. Joseph, Mich.: American Society of Agricultural Engineers.

Monath, T. P., C. B. Cropp, and A. K. Harrison. 1983. Mode of entry of a neurotropic arbovirus into the central nervous system: Reinvestigation of an old controversy. *Laboratory Investigations* 48: 399–410.

Murua, V. S., and V. A. Molina. 1990. Desipramine and restraint stress induce odor conditioned aversion in rats: Suppression by repeated conditioning. *Psychopharmacology* (Berl) 102: 503–506.

Nance, D. M., D. Rayson, and R. I. Carr. 1987. The effects of lesions in the lateral septal and hippocampal areas on the humoral immune response of adult female rats. *Brain Behavior Immunology* 1: 292–305.

Neutra, R., J. Lipscomb, K. Satin, and D. Shusterman. 1991. Hypotheses to explain the higher symptom rates observed around hazardous waste sites. *Environmental Health Perspectives* 94: 31–38.

Nielsen, V. C., J. H. Voorburg, and P. L'Hermite, eds. 1991. *Odour prevention and control of organic sludge and livestock farming,* 1–391. London: Elsevier Applied Science, 1986. *Odour and ammonia emissions from livestock farming,* 1–222. London: Elsevier Applied Science.

Nordstedt, R. A., and E. P. Taiganides. 1971. "Meteorological Control of Malodors from Land Spreading of Livestock Wastes." In *Livestock Waste Management and Pollution Abatement,* 107–116. St. Joseph, Mich.: American Society of Agricultural Engineers.

O'Leary A. 1990. Stress, emotion, and human immune function. *Psychology Bulletin* 108: 363–382.

Ono, T. P. G. Luiten, H. Nishijo, M. Fukuda, and H. Nishino. 1985. Topographic organization of projections from the amygdala to the hypothalamus of the rat. *Neuroscience Research* 2: 221–238.

Oomura, Y. "Input-Output Organization in the Hypothalamus Relating to Food Intake Behavior." In *Handbook of the Hypothalamus.* vol. 2, ed. P. Morgane and J. Panksepp, 557–620. New York: Marcel Dekker, 1980.

———. "Modulation of Prefrontal and Hypothalamic Activity by Chemical Senses in the Chronic Monkey." In *Umami: A Basic Taste,* ed. Y. Kawamura and M. R. Kare, 481–509. New York: Marcel Dekker, 1987.

Oomura, Y., T. Ono, and H. Ooyama. 1970. Inhibitory action of the amygdala on the lateral hypothalamic area in rats. *Nature* 228: 1108–1110.

Pain, B. F., C. R. Clarkson, V. R. Phillips, J. V. Klarenbeek, T. H. Misselbrook, and M. Bruins. 1991. Odour emission arising from application of livestock slurries on land: Measurements following spreading using a micrometeorological technique and olfactometry. *Journal of Agricultural Engineering Research* 48: 101–110.

Phillips, M., and J. Greenberg. 1992. Ion-trap detection of volatile organic compounds in alveolar breath. *Clinical Chemistry* 38: 60–65.

Rains, B. A., M. J. Deprimo, and I. L. Groseclose. 1973. *Odors emitted from raw and digested sewage sludge. Environmental protection technology series.* Prepared for the Office of Research and Development. Environmental Protection Agency. Washington, D.C.: U. S. Government Printing Office.

Raymer, J. H., E. D. Pellizzari, K. W. Thomas, and S. D. Cooper. 1991. Elimination of volatile organic compounds in breath after exposure to occupational and environmental microenvironments. *Journal of Exposure, Analysis, and Care in Environmental Epidemiology* 1: 439–451.

Ritter, W. F. 1989. Odour control of livestock wastes: State-of-the-art in North America. *Journal of Agricultural Engineering Research* 42: 51–62.

Roberts, E. 1986. Alzheimer's disease may begin in the nose and may be caused by aluminosilicates. *The Neurobiology Aging* 7: 561–567.

Roszman, T. L., R. J. Cross, W. H. Brooks, and W. R. Markesbery. 1982. Hypothalamic-immune interactions. II. The effect of hypothalamic lesions on the ability of adherent spleen cells to limit lymphocyte blastogenesis. *Immunology* 45: 737–742.

Rotton, J. 1983. Affective and cognitive consequences of malodorous pollution. *Basic Applied Social Psychology* 4: 171–191.

Saphier, D., O. Abramsky, G. Mor, and H. Ovadia. 1987. Multiunit electrical activity in conscious rats during an immune response. *Brain Behavior Immunology* 1: 40–51.

Satin, K. P., S. Huie, and L. Croen. 1986. *Operating industries inc. health effects study.* Berkeley, Calif.: California Department of Health Services, Epidemiological Studies Section.

Satin, K. P., G. Windham, J. Stratton, and R. Neutra. 1987. *Del Amo-Montrose Health Effects Study.* Berkeley: California Department of Health Services, Epidemiological Studies Section.

Scarborough, M. E., R. G. Ames, M. J. Lipsett, and R. J. Jackson. 1989. Acute health effects of community exposure to cotton defoliants. *Archives of Environmental Health* 44: 355–360.

Schiffman, S. S. 1983. Taste and smell in disease. *New England Journal of Medicine* 308: 1275–1279; 1337–1343.

Schiffman, S. S., E. A. Sattely-Miller, M. S. Suggs, and B. G. Graham. 1994. The effect of pleasant odors and hormone status on mood of women at midlife. *Brain Research Bulletin* 36: 19–29.

Schiffman, S. S., M. S. Suggs, and E. A. Sattely-Miller. 1994. Effect of pleasant odors on mood of males at midlife: Comparison of African-American and European-American men. *Brain Research Bulletin* 36: 31–37.

Scott, J. W., and C. Pfaffmann. 1967. Olfactory input to the hypothalamus: Electrophysiological evidence. *Science* 158: 1592–1594.

Shipley, M. T. 1985. Transport of molecules from nose to brain: Transneuronal anterograde and retrograde labeling in the rat olfactory system by wheat germ agglutinin-horseradish peroxidase applied to the nasal epithelium. *Brain Research Bulletin* 15: 129–142.

Shusterman D. 1992. Critical review: The health significance of environmental odor pollution. *Archives of Environmental* 47: 76–87.

Skarp, S. 1975. "Manure Gases and Air Currents in Livestock Housing." In *Managing Livestock Wastes,* 362–365. St. Joseph, Mich.: American Society of Agricultural Engineers.

Sneath, R. W. 1988. The effects of removing solids from aerobically treated piggery slurry on the VFA levels during storage. *Biological Wastes* 26: 175–188.

Sneath, R. W., C. H. Burton, and A. G. Williams.1992. Continuous aerobic treatment of piggery slurry for odour control scaled up to a farm-size unit. *Journal of Agricultural Engineering Research* 53: 81–92.

Sneath, R. W., and A. G. Williams. 1990. The possible importance of wind aeration in controlling odous from piggery slurry stored after aerobic treatment. *Biological Wastes* 33: 151–159.

Solomon, G. F. 1987. Psychoneuroimmunology: Interactions between central nervous system and immune system. *Journal of Neuroscience Research* 18: 1–9.

Solvason, H. B., V. K. Ghanta, J. F. Lorden, S. J. Soong, and R. N. Hiramoto. 1991. A behavioral augmentation of natural immunity: Odor specificity supports a Pavlovian conditioning model. *International Journal of Neuroscience* 61: 277–288.

Solvason, H. B., V. K. Ghanta, S. J. Soong, C. F. Rogers, C. M. Hsueh, N. S. Hiramoto, and R. N. Hiramoto. 1992. A simple, single, trial-learning paradigm for conditioned increase in natural killer cell activity. *Proceedings of the Society of Experimental Biological Medicine* 199: 199–203.

Stone, A. A., D. S. Cox, H. Valdimarsdottir, L. Jandorf, and J. M. Neale. 1987. Evidence that secretory IgA antibody is associated with daily mood. *Journal of Personality and Social Psychology* 52: 988–993.

Sutin, J., R. L. McBride, R. H. Thalmann, and E. L. Van Atta. 1975. Organization of some brainstem and limbic connections of the hypothalamus. *Pharmacology and Biochemical Behavior* 3(1 Suppl.): 49–59.

Sweeten, J. R., and J. R. Miner. 1993. Odor intensities at cattle feedlots in nuisance litigation. *Bioresource Technology* 45: 177–188.

Takagi, S. F. 1986. Studies of the olfactory system in the old world monkey. *Progress in Neurobiology* 27: 195–250.

Tanabe, T., M. Iino, and S. F. Takagi. 1975. Discrimination of odors in olfactory bulb, pyriform-amygdaloid areas, and orbitofrontal cortex of the monkey. *Journal of Neurophysiology* 38: 1284–1296.

Tazawa, Y., N. Onoda, and S. F. Takagi. 1987. Olfactory input to the lateral hypothalamus of the old world monkey. *Neuroscience Research* 4: 357–375.

Thomas, K. W., E. D. Pellizzari, and S. D. Cooper. 1991. A canister-based method for collection and GC/MS analysis of volatile organic compounds in human breath. *Journal of Analytic Toxicology* 15: 54–59.

Tyrey, L., and A. V. Nalbandov. 1972. Influence of anterior hypothalamic lesions on circulating antibody titers in the rat. *American Journal of Physiology* 222: 179–185.

von Kluge, S., and F. R. Brush. 1992. Conditioned taste and taste-potentiated odor aversions in the Syracuse high-and low-avoidance (SHA/Bru and SLA/Bru) strains of rats (*Rattus norvegicus*). *Journal of Comparative Psychology* 106: 248–253.

Waggoner, M. "Tobacco Farmers Going Hog Wild for Pigs," *The Herald-Sun,* Durham, N.C., 8 August 1994.

Wallace, L., W. Nelson, R. Ziegenfus, E. Pellizzari, L. Michael, R. Whitmore, H. Zelon, T. Hartwell, R. Perritt, and D. Westerdahl. 1991. The Los Angeles TEAM study: Personal exposures, indoor-outdoor air concentrations, and breath concentrations of 25 volatile organic compounds. *Journal of Exposure Analysis Care and Environmental Epidemiology* 1: 157–192.

Walsh, E. "Corporate Pork Worries Iowa Family Farmers," *The Washington Post,* 19 October 1994, A1, A14.

Warburton, D. J., J. N. Scarborough, D. L. Day, A. J. Muehling, S. E. Curtis, and A. H. Jensen. 1980. "Evaluation of Commercial Products for Odor Control and Solids Reduction of Liquid Swine Manure." In *Livestock waste: A renewable resource,* 309–313. St. Joseph, Mich.: American Society of Agricultural Engineers.

Warner, P. O., K. S. Sidhu, and L. Chadzynski. 1990. Measurement and impact of agricultural odors from a large scale swine production farm. *Veterinary and Human Toxicology* 32(4): 319–323.

Weisse C. S. 1992. Depression and immunocompetence: A review of the literature. *Psychological Bulletin* 3: 475–489.

Wilcosky, T. C. 1993. Biological markers of intermediate outcomes in studies of indoor air and other complex mixtures. *Environmental Health Perspective* 101(Suppl 4): 193–197.

Williams, A. G., M. Shaw, C. M. Selviah, and R. J. Cumby. 1989. The oxygen requirements for deodorizing and stabilizing pig slurry by aerobic treatment. *Journal of Agricultural Engineering Research* 43: 291–311.

Winneke, G., and J. Kastka. 1977. *Odor pollution and odor annoyance reactions in industrial areas of the Rhine-Ruhr region,* 471–479. Oxford: IRL Press. Olfaction and Taste VI, Paris.

Chapter 6

Large-Scale Swine Production and Water Quality

Laura L. Jackson

There is an old joke about an Iowa farmer and an Arkansas farmer comparing notes on the hog business. The Iowa farmer says, proudly, that it takes him six months to raise a hog from birth to market. The Arkansas farmer expresses admiration and then admits it takes him two years. "But then," he says, "what's time to a hog?"

There is significant ecological and historical meaning to this joke. At one time, agricultural production in a region was set by ecological limitations—the amount and timing of sunshine and rainfall during the growing season and the depth, fertility and resiliency of the soils. Up until about thirty years ago, farmers had no option but to accept their ecological limits.

Today the speed at which animals can be raised to maturity appears to bear little relationship to actual ecological conditions. Farmers can import feed, fertilize crops, and air condition production buildings to achieve the same rate of production virtually anywhere. The joke really reflects a misunderstanding between generations, between the vast majority of people who historically adapted with contemporary energy—sunlight, crops, wood—and the world of today, which runs on a fuel stored millions of years ago.

Large-scale, high density confinement of livestock is the latest development in a century-long campaign to overcome ecological constraints through agricultural industrialization. What are the ecological consequences of raising more hogs, faster, on less land than ever before? First I review several known and suspected consequences of large-scale swine production on water quality. Then I examine the reasons these well-documented concerns are not reflected in current public policy. I argue that we lack critical data on complex and variable biological, geological, social, and farming systems and their interactions; that the burden for proving harm is wrongly placed and the standard of proof is too high; that the framework for analyzing agricultural pollution is too narrow; and that scientists and policy makers accept high capital, high technology innovation as inevitable, without the benefit of meaningful, scientific comparison to alternative systems.

LIVESTOCK WASTE AND WATER QUALITY

Figure 6.1 summarizes the various concerns raised over the impacts of large-scale swine production on water quality. Rural citizens have asked a number of important questions about the impact of large-scale swine production on water quality (Thu 1996): Why do livestock facilities not treat their wastes as cities do? Do the lagoons and pits leak? Are larger or older lagoons more likely to leak than new ones? Are large, corporate-owned hog confinements more likely to cause water pollution than smaller operations? Will the spreading of manure on fields cause excess nutrients to leach into the groundwater or run off into streams?

A number of other questions are not being asked by citizens but are of concern in the scientific community, especially in countries with high density livestock production (Jackson et al. 1996): What happens to the surrounding environment when ammonia vaporizes (when ammonia becomes a gas)? Can swine facilities contaminate drinking water with disease agents that affect people (human pathogens) or spread antibiotic resistance among bacteria? Will

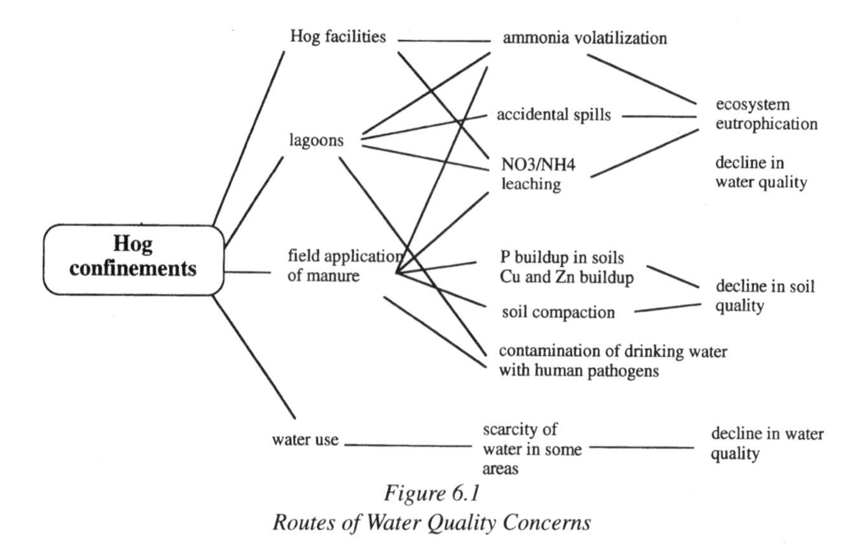

Figure 6.1
Routes of Water Quality Concerns

pollution from fields increase due to replacement of mixed crop and livestock farms with cash grain enterprises?

Manure Treatment and Storage

One hog generates up to eight times as much solid waste as a person. While small towns are required to treat their waste before returning it to a river or stream, livestock facilities housing several thousand animals are not. This is because swine and other animal wastes are applied to land, not discharged into the rivers—at least not legally. Because city wastewater treatment facilities discharge effluent directly into surface waters, they must eliminate human pathogens and greatly reduce the amount of oxygen bacteria take from the water. Biochemical oxygen demand (BOD) is the oxygen demand created when heterotrophic miocrorganisms feed on carbon and nutrients in raw sewage. High BOD reduces the oxygen available in the water for aquatic life. BOD and human pathogens are greatly reduced by aerobic decomposition of wastes. Swine waste, in contrast, is applied to land where sunlight and microbial activity in the soil generally reduce pathogen populations, and the nutrients are used by crops. BOD is not a problem in soil that is adequately exposed to air.

Prior to confined livestock facilities, hogs were largely raised outdoors. Sows farrowed in individual, widely spaced huts and grazed in pastures while little pigs were still nursing (six to eight weeks). Manure from farrowing sows and pigs was deposited directly on pastures which were rotated with row crops. The nitrogen from legumes in the pasture and nutrients in the hog wastes fertilized the row crops. Weaned pigs were fed on concrete or mud lots that sloped down from a partially covered shelter. A hog's natural tendency is to defecate as far from its shelter as possible, so manure was concentrated at the far end of the feed lot. A shovel or front-end loader on a tractor was used to scrape the dry manure out and load it into a manure spreader to haul it to the fields.

Two liquid-based methods now prevail over the solid-based disposal system. Both systems raise pigs in confined quarters on concrete slats over concrete-lined manure pits. Manure and other wastes drop through the slats into pits below where they accumulate in a highly concentrated form directly underneath the animals until the pit is pumped out. The more popular system involves the use of an earthen, outdoor storage lagoon. In this system, wastes in the pits underneath the pigs are flushed periodically with fresh or recycled lagoon water and pumped into the lagoon. Lagoon waste is more diluted than pit storage waste. In both systems, bacteria quickly use up the available oxygen. Bacteria that can survive without oxygen (anaerobic) continue to break down the wastes.

In both systems, large quantities of hydrogen sulfide gas and ammonia, in addition to many other gaseous compounds, are released as Kelley Donham describes in his chapter. Fewer odors are reported from pit systems because they are enclosed by the swine facility and vented from narrow openings, while open air lagoons outside can release plumes of gas as wide as the lagoon itself. While lagoons are currently more common, pits are becoming more popular in some areas because they tend to reduce odors.

Lagoon Performance

Animal waste lagoons pose risks to both surface water and groundwater. Above-ground containment dams, or berms, may collapse and spill some or all of their contents which can flow downhill toward the nearest stream. Careless maintenance of lagoon walls and overfilling, especially with above-average rainfall, can weaken the walls, causing damage ranging from "minor" spillovers to complete breaches of the lagoon structure. The latter happened in North Carolina in June 1995, when 22 million gallons of swine waste spilled into a tributary of the New River, killing fish and shellfish along nineteen miles of rich estuarial habitat, destroying cropland in its path. In 1996 the fish and other organisms in the Iowa River were killed along a ten-mile stretch by a 1.5 million gallon manure lagoon spill. The spill originated from a swine waste lagoon built on top of an old field drainage tile system. One of the tiles traversed the lagoon wall and became a direct conduit to the river. Other kinds of problems have occurred such as when an operator reduced his lagoon level by pumping waste from the lagoon into a nearby grass waterway and forgot to turn off the pump.

While recently built lagoons are designed with enough capacity to withstand the record twenty-five-year, twenty-four-hour rainfall, weather still can contribute to lagoon breaches (Jackson et al. 1996). Several consecutive days of rainfall can prevent the operator from pumping out the lagoons onto saturated fields, while berms weaken and levels continue to rise in the already full lagoon.

Less dramatic but more worrisome are lagoons that leak into the groundwater. All newly constructed earthen lagoons leak until soil pores in the walls and floor are gradually sealed with solids from animal waste. Early studies showed almost no seepage after a few months in operation (Davis, Fairbank, and Weisheit 1973; Chang 1974). However, Huffman and Westerman (1995) studied the seepage rate and total nitrogen export of unlined lagoons in North Carolina across a range of soil types. Seepage rates were related to soil texture, with coarse soils allowing more seepage. In their study, 55 percent of the lagoons experienced "moderate to severe" rates of seepage. Pockets of sandy material in otherwise sound lagoon walls or floors could result in serious seepage.

Regardless of lagoon construction, each time the lagoon is pumped out the walls dry out and crack. As the lagoon refills, a pulse of waste seeps out of these cracks until they swell and close. If the lagoon is refilled slowly, moisture can reach the walls through capillary action and minimize seepage (Midwest Plan Service 1985).

Much emphasis has been placed on proper lagoon design, especially the installation of clay liners on the bottom and sides. Liners make the soil pores smaller so wastes can fill in the remaining pores and minimize further seepage. This is particularly important in regions with sandy soils. However, with or without liners, lagoons are exposed to physical soil-forming forces such as freezing and thawing, burrowing animals, plant roots, and cracking caused by drying walls when the lagoon is pumped out. McCurdy and McSweeney (1993) examined the lagoon liner at the surface, middle, and base of a leaking ten-year-old Wisconsin dairy sewage lagoon and found earthworm holes and plant root channels going through the liner at all three depths.

Lagoon liners can also be disrupted in other ways. Improper dredging of sludge off the bottom of the lagoon, which typically occurs every five to fifteen years, can reduce the thickness of the liner or break through it entirely. Erosion of lagoon berms may reduce or eliminate the liner at the top. The Iowa Department of Natural Resources and the Geological Survey Bureau recently inspected lagoons owned by A. Jack DeCoster in north-central Iowa and found small-scale erosion in all fourteen units observed and gully erosion or "extensive" erosion in eleven of the fourteen lagoons. At one site, a gully 1.5 feet deep occurred in one wall (Iowa DNR 1996). Because of the dynamic and universal processes of soil erosion, earthworm and plant root activity, plus routine pumping and refilling of lagoons, and periodic dredging, it is clear that maintenance and management of lagoons are critical in determining their actual performance. This suggests the need for unannounced inspections of lagoons by regulatory agencies to assess lagoon performance, in addition to providing engineering specifications for lagoon design which only assume their proper performance.

Lagoons are not routinely inspected by government regulators, and leaks are not always easy to detect. Different parts of the lagoon or just one small section may be seeping while the rest of the structure is sound. Multiple groundwater testing wells placed around the lagoon at varying distances and depths are required. The volume and direction of groundwater flow determines the degree to which the seepage is diluted. A literature survey of scientific research by Jackson and colleagues (1996) found no study in which the performance of lagoons is related to age, size, ownership, management practice, or design. Because we lack such studies it is not possible to assess the extent of groundwater contamination from manure lagoons in any region of the country.

Leaks to the Atmosphere

Despite the well-publicized concern to prevent swine waste from reaching surface waters and groundwaters, lagoons routinely leak to the atmosphere. As much as 70 to 80 percent of the lagoon's nitrogen changes from a liquid to a gas ("ammonia volatilization") and escapes into the atmosphere (Sutton 1994).

In contrast, dry manure handling systems lose 15 to 40 percent of their nitrogen to the atmosphere. The gaseous ammonia quickly returns to earth, stripped from the atmosphere by rain or trapped by trees, grass, or water bodies. This process is called "atmospheric deposition." In the Netherlands, atmospheric deposition of nitrogen is ten times greater than natural levels, and the greatest deposition (fifty to sixty-five kg nitrogen per hectare per year) occurs in the southeastern part of the country where the livestock industry is the most intensive and concentrated (references in Berendse, Aerts, and Bobbink et al. 1993).

Nitrogen-enriched rainfall can damage natural habitats substantially. In Dutch heathlands, the dominant dwarf shrub *Erica tetralix* has been almost totally replaced by the grass *Molinia caerulea* (Berendse, Aerts, and Bobbink 1993), and several rare species have disappeared. Fifty species that grow in low to medium nutrient conditions in dune systems have declined while those that require more nitrogen now dominate (Verhoeven, Kemmers, and Koerselman 1993). The same is true in other parts of the Netherlands, central Europe, Great Britain, and southern Sweden (Marrs 1993). In the upper Midwest, an eleven-year study by Inouye and Tilman (1995) showed that even minor increases in annual nitrogen fertilization (as little as 0.34 kg per hectare) caused major shifts in species composition, including a reduction in total species diversity. Prairies in the Midwest may be particularly susceptible to ecological shifts due to excess nitrogen fertilization.

Manure Spreading

Manure spreading is an ancient practice that replenishes soil nutrients and can contribute to soil texture, organic matter, and overall soil health. However, if done improperly it can kill fish and pollute both ground and surface waters. In Iowa from 1991 to 1993, four of six reported fish kills (more than 47,000 fish killed) whose cause could be determined were attributed to livestock feedlots (Iowa Department of Natural Resources 1994). Runoff is more likely where the soil has a slow absorption rate, such as frozen or snow-covered ground and clay soils (Schwab et al. 1993).

There is also a risk of groundwater contamination by nitrates. In contrast to ammonia and phosphorus compounds, nitrates are water soluble and move easily through the soil. Nitrate leaching to the groundwater is most likely when: (a) soil nitrogen is already high; and (b) when the amount of nitrogen applied exceeds the needs of the crop. The higher the nitrate concentrations, the greater the losses to shallow groundwater (Baker and Johnson 1981; Kanwar, Johnson, and Baker 1983). As nitrogen application increases, corn grown for silage uses a diminishing proportion (forty percent to eighty percent) of the nutrients applied (Antoun et al. 1985). Overapplication of nitrogen in any form is likely to cause nitrate leaching. Many studies have linked nitrate concentrations in groundwater to agricultural land use (Ritter and Chirnside 1984; Hallberg and Keeney 1993). However, there is no easy way to trace the source of nitrogen in nitrate-contaminated groundwater.

A number of factors complicate safe, precise application of manure for fertilizer. Even distribution of liquid animal waste is difficult, and its nutrient content is heterogeneous. Each load should be tested to determine the proper rate of application. There is a strong economic incentive to apply liquid wastes as close as possible to the waste source rather than where nutrients are needed, because their transport is a significant expense. For optimum nutrient uptake, nutrients should be applied at or near the time of crop demand. This may not be possible (as in the case of row crops) or it may not coincide with the need to empty a lagoon, particularly in fall or winter. In regions where swine waste is customarily applied to corn fields, the availability of abundant and free animal waste may discourage corn growers from rotating with leguminous crops such as alfalfa or soybeans. Even if manure could be applied as precisely as inorganic fertilizer, there would still be a groundwater problem in some areas. Nitrate leaching that exceeds the EPA's drinking water standard has occurred where corn is produced with the economic optimum amount of chemical fertilizer (Roth and Fox 1990).

The relative nutrient content of animal waste often does not match the needs of a crop. There is about 1.5 parts nitrogen (N) for every phosphorus part (P) in hog waste. Growing corn plants need a nitrogen to phosphorous ratio of 6–1 (Pennsylvania State University 1994). Consequently, when waste is applied to satisfy the nitrogen demands of the crop, phosphorus builds up. In the Netherlands, which has a long history of concentrated animal agriculture, about 1 million acres of phosphorous saturated soils has resulted in significant contamination of surface and ground waters, and farmers must drive long distances to dispose of animal wastes (references in Vos and Opdam 1993).

Pathogens

Several organisms that cause human disease have been found in swine feces and lagoons, and some of them may survive in lagoons and spread to drinking water supplies via leaks or leaching from fields to which the waste is applied. Up to 50 percent of bacteria and 90 percent of viruses may be destroyed in lagoons (Hurst 1995). Survival after the wastes are spread on land depends on the temperature, moisture, pH, sunlight, competition with other microorganisms, and predation (Bitton and Harvey 1992).

Gastrointestinal illness resulting from consumption of contaminated water continues to be a threat to human health, especially in rural areas where water supplies may not be tested regularly. The source of pathogens responsible for outbreaks of gastrointestinal illness often is never identified. Currently there are no known methods for tracing microbial contamination to swine wastes. Furthermore, many potential pathogens are difficult or impossible to grow in the lab and thus cannot be identified. New molecular techniques may enable us to find and identify potential pathogens. However, the application of these methods to environmental and public health has been slow, and most of the work is being done in countries outside the United States (references in Jackson et al. 1996). Bacteria in swine waste may also carry genes for broad spectrum antibiotic resistance. Hogs in confinement are routinely fed antibiotics, though not those normally given to humans, and persistent use of antibiotics leads to evolution of resistance in bacterial populations. Genes for resistance to a particular antibiotic can readily be transferred to other bacteria.

TRANSLATING SCIENCE INTO PUBLIC POLICY

Given the list of known and suspected problems associated with large-scale concentrated swine production, how has public policy reflected this knowledge? Why has society effectively given carte blanche permission for this approach to swine production?

Proving Harm

First, our current approach to technology is to assume it is innocent until proven guilty. The burden is on the public to prove harm, rather than on industry to demonstrate safety. It is difficult to demonstrate harm when data are scarce. For instance, Huffman and Westerman's (1995) survey of eleven lagoons on different soil types is the only published survey to date of lagoon performance that specifically addresses the question of soil type. There appears to have been no field surveys of lagoon performance related to lagoon age, size, management practice, or ownership. Similarly, no data have been collected to assess the actual water quality impacts of different swine operations varying in their methods of field application of waste. We do not know whether manure management plans accepted by regulatory agencies prevent pollution because there are no data.

The lack of data is also a problem in documenting the health effects of contaminated drinking water. Infantile methemoglobinemia, a potentially fatal disease caused by high nitrate levels in drinking water, is perceived to be quite rare, even though at least 15 percent of rural wells in the Midwest exceed the U.S. EPA drinking water standard of 10 milligrams per liter for nitrates. Meyer (1996) found that this perception may be incorrect due to a lack of research since 1950. The few reports that have been published suggest that the disease is likely to pose a significant health risk to formula-fed infants (Meyer 1996).

Sometimes harm is difficult to prove because natural systems are complex. For example, the long lag time between water contamination at one source and detection at a well and the limited ability to map groundwater movement make it difficult to pinpoint which livestock facility or other pollution source is responsible for contamination. A second example involves disease organisms. We know little about the ability of human pathogens to survive in hog waste and eventually be transmitted to drinking water supplies. Atmospheric deposition of nitrogen is a third example. Due to extensive mixing of the atmosphere and complex weather patterns, it is impossible to trace nitrogen-enhanced rainfall to particular sources, or even to animal waste lagoons in general.

Groundwater and surface water contamination caused by livestock operations are almost certainly underreported, in part because the evidence for pollution must be exceedingly strong. For example, a rural resident in an area of many large swine facilities reported a fish kill in a nearby stream in mid-October 1996. Before the kill, six farmers, some large-scale operators and some not, upstream from the kill spread manure adjacent to the stream. A couple of days later there was a one-inch rain, and the next day the fish were dead. Despite timely detection and reporting of the fish kill, the county game warden was not able to positively identify the source of the fish kill. In order to assign a cause, he explained that it was necessary either to witness runoff coming from a field, or to measure high levels of livestock manure in the water. The cause of the fish kill was therefore listed as "unknown." About half of reported fish kills in Iowa are of unidentified cause (Iowa DNR 1994). If the land in the vicinity of large-scale livestock facilities loses rural residents due to the farm economics that John Ikerd describes in his chapter and the odor that Susan Schiffman describes in her chapter, then fewer wells will be tested, fewer streams will be inspected by residents, and even more cases of contamination will go unreported.

A related problem with the high standard of proof described above is that there are few clear benchmarks for what constitutes an unacceptable level of water degradation in a region. Most water sources in upper Midwest farming regions are already far from pristine, as farmer Blaine Nickles details from first-hand experience in his chapter. Currently no river or stream in Iowa supports all of its designated potential uses (Iowa DNR 1994), and many are unusable for drinking or bathing. Water quality degradation has been and will continue to be gradual, it will vary from place to place; and it will be the result of multiple causes, including small, diversified farms, soil erosion, private sewage systems, and large-scale swine operations. As long as there is no political will to draw clear lines specifying how much is too much, decision makers will err on the side of contamination rather than caution.

Framing the Questions

A second reason for the lack of public policy response to questions of environmental safety is that we have not framed the issues broadly enough. Most discussions of the impact of large-scale swine operations on water quality address the immediate question: Where will the manure go? But this is only part of the picture (figure 6.1) which should include the entire food chain (figure 6.2).

The corporate-owned hog operation is predicated upon the cash grain farm. The proliferation of cash grain farms and loss of mixed crop-livestock operations has predictable ecological consequences:

- Cash grain farms rely primarily on imported rather than local biologically fixed nitrogen; this leads to the greater potential for fertilizer nitrogen leaching.
- Cash grain farms require annual application of pesticides to control pests and weeds. Traditional crop rotations involving oats or wheat and two years of mixed hayfields manage pests and weeds biologically (Liebman and Dyck 1993).
- Cash grain farms expose 100% of their crop ground to the elements each year, increasing the potential for soil erosion (National Research Council 1993).
- Low or no-till methods can be used to reduce soil erosion, but they increase herbicide and pesticide dependence which in turn can exacerbate some soil quality problems and cause a shift in problem weed species.

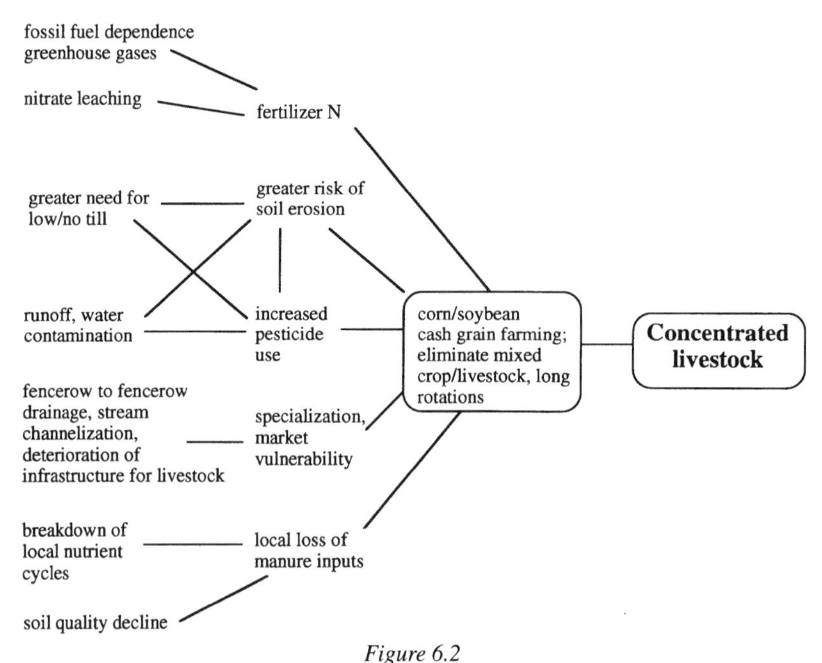

Figure 6.2
A Broadened Framework for Understanding Concentrated Livestock Production

- Cash grain systems are typically much larger than the mixed livestock-grain-pasture systems of the past. The combination of specialization, larger operation size, specific farm policies, and high interest rates has been associated with more thorough exploitation of available land. Conservation acres, hedgerows, wetlands, and woodlands all became more scarce during the 1970s as mixed crop-livestock farms shifted increasingly to cash grain operations. These marginal lands once protected water resources and provided habitat for wild plants and animals (Warner 1994; Jackson, in press).
- Separation of feed production from animal production means using more fossil fuel. Therefore, this system probably contributes more to carbon dioxide emissions, a greenhouse gas (see figure 6.2).

The fact that land use typically has not been included in the large-scale hog operation debate is evidence that a large part of the water cycle—namely the way the landscape processes rainfall—has been left out of the equation. This shortcoming is revealed in the curious expression *nonpoint source pollution,* which would be better called "landuse pollution" in order to correctly assign responsibility for the problem.

The consequences of land use change for the water cycle have been well documented. In 1988, 30 percent of the nation's stream and river miles could not support fishing and swimming. Agriculture (nonpoint source pollution) accounted for 55 percent of the problem of those river and stream miles (Doppelt et al. 1993). In the Big Springs, Iowa, study, groundwater nitrate concentrations tripled from about three milligrams per liter in the late 1950s to ten milligrams per liter in 1983. During the same time period, fertilizer nitrogen application rose 300 percent and application of manure nitrogen rose 30 percent (Hallberg et al. 1984).

To date, land use change, not manure spreading per se, is responsible for groundwater contamination. Prior to 1950, manure and biological nitrogen were the most important sources of nitrogen for crop production. Manure spreading was a universal practice, yet average annual groundwater nitrate levels were below three milligrams per liter (Halberg et al. 1984). It does not matter whether groundwater contamination comes from crop production (fertilizer nitrogen) or from animal wastes (manure nitrogen). The entire production system results in groundwater contamination.

The language of the Clean Water Act of 1973 established a broad definition of surface water quality, including "chemical, physical, and biological integrity" (Doppelt et al. 1993). In practice, however, monitoring and regulation of water quality at the state level have been limited mainly to chemical and physical criteria. These criteria can fail to detect many water quality violations that biological measures can detect. In a recent study in Ohio monitoring 645 waterbody segments, biological impairment of aquatic life was evident in almost 50 percent of the cases in which there were no violations of chemical water quality criteria (Doppelt et al. 1993). A broader framework for the protection of water quality would have provided far more sophisticated and sensitive protection of water from farming practices, including livestock production.

A final example of an inadequate framework of understanding is our limited attention to the regional nutrient balance. On a large scale, the environmental effects of the spread of large-scale hog operations depend more on the net nutrient and carbon balance for the region and less on individual management decisions on particular farms. An excess of nutrients entering the region in the form of fertilizer and feed, over those exiting the region in the form of feed, meat, or other products will place stress on the ecosystem and problems will be inevitable. This already has occurred in the Netherlands, where 80 percent of livestock feed is imported and water is contaminated with excess phosphorus and nitrogen (Vos and Opdam 1993). Improved management and technology can mitigate but not "solve" waste problems.

The consequences of adding more nitrogen to a system than we take away can be felt as far away as the Gulf of Mexico. The nitrogen in the water at the mouth of the Mississippi is now estimated to be two to three times greater than it was before European settlement, mostly due to fertilization and mineralization of stored nutrients in the soils of the Mississippi River basin. Nutrient-fed algal blooms in the Gulf die and sink to the bottom, using up oxygen in the decay process. This contributes to a spreading "dead zone" of low oxygen (hypoxic) waters (Sparks 1995). This reduces populations of marine animals such as shrimp and oysters and changes the ecology of bays and estuaries. Artificial fertilization of aquatic and terrestrial systems by human-generated nitrogen (eutrophication) is a global change issue on par with climate change and deforestation. Humans are responsible for more than doubling the earth's annual nitrogen budget, most of which has occurred since 1960 (Vitousek 1994).

The debate over large-scale swine confinement operations rarely touches on the Gulf of Mexico because few are willing to draw the boundaries of consideration so broadly. We continue to focus narrowly on the specifics of lagoon design and manure management without stepping back to acknowledge the bigger picture. Meanwhile, we will not solve the water quality problem in the Gulf of Mexico without addressing fertilizer use, cropping practices, and livestock in the Midwest.

If we decide without due consideration that the environmental problems large-scale livestock systems cause are no more severe than the problems caused by small, dispersed systems, and we are wrong, we will have committed a serious, potentially irreversible error. Mapstone (1995) has studied scientific error and identified situations in which we wrongly conclude that two things are *not* different from one another when in fact they are. This is most likely to happen when: (1) we do not specify what magnitude of impact is considered important; (2) we set an extremely high standard for showing two things are different and collect inadequate data; or (3) we fail to ask the right questions. All three conditions are met in the case of large-scale hog operations.

WHAT SHOULD WE DO?

Citizens, scientists and policy makers should better understand the water quality impacts of large-scale swine confinements. First, we should insist on better data. It is invalid to make assumptions about the integrity of structures such as manure lagoons without actual performance data. Likewise, it is invalid to assume that manure management plans are actually followed, or that when they are followed they avoid manure runoff or leaching, without performance data. We need field surveys of actual performance, over a range of climates and soil types and a range of management and ownership systems. This applies to smaller, less capital-intensive systems as well. Where the scale and complexity are too great to make definite conclusions, prudent policy should be designed with the conservative assumption that what we do not know *will* hurt us.

Second, we must enlarge the analysis of livestock production and water quality to include how the feed was raised and take into consideration the entire water cycle. So-called nonpoint source pollution caused by cropping practices is a major threat to water quality and from an ecosystem point of view, crops and livestock cannot be separated. We must also pay attention to the mass balance in various regions—total nutrients in versus total nutrients out—and fully acknowledge the long-term consequences of nutrient excesses or deficits.

Third, scientists must examine the assumption that high capital, high technology innovations in farming, such as large-scale swine operations, are inevitable. Many scientists naively assume that family farms are competing on a level playing field with factory farms and that the rules of the game are "management skill, technology, efficiency and profitability" (Strange 1989). With this myth in place, it is easy to accept industry claims that large-scale hog operations are an inevitable consequence of economic Darwinism and that the best survive. If large-scale livestock confinements are better and are going to prevail, many agricultural scientists reason that it is not necessary to study other types of livestock systems.

As a result, large amounts of research and research funding are spent accommodating, fine-tuning and promoting large-scale hog confinements without meaningful comparison to other farming systems. It is important for scientists and policy makers to recognize that scientific knowledge is created within a social and political context that shapes the outcomes of investigation by determining the questions that are asked, which new knowledge is accepted, and which is ignored. The more conscious scientists become of this process, the more likely we are to do truly independent and unbiased research.

My fourth recommendation is that we insist upon a meaningful comparison of alternative swine production systems. Large-scale hog confinements take the existing system of increasingly input-intensive hog production and merely increase its magnitude, while the operating logic has changed little. As a result, the current debate over large-scale hog confinements is limited to two very similar options. Many profitable farms use alternative management systems such as crop rotation integrated with livestock production to manage simultaneously weeds, pests, animal pathogens, soil fertility, and health (Liebman and Dyck 1993). On one farm in northeast Iowa, hogs are farrowed outdoors three seasons of the year and rotated from pasture to pasture. Strips of corn next to the pastures give the pigs needed shade, but an electric fence keeps the mothers out until fall when they are allowed to move in and harvest the corn. "Hogging down" the corn can save up to fifty dollars an acre in harvesting and drying costs. Every year the strips of clover pasture, corn, and oats are rotated to provide nutrients and reduce weeds and diseases. The farmer planted rows of hybrid poplars, chestnuts, and hazelnuts between each of the corn-oats-pasture units to control erosion, provide shade for the pigs, and eventually give him nut crops. Direct-marketed chickens roam around after the sows, cleaning up what they fail to eat and going through feces for undigested morsels. The finish hogs are kept in a deep-bedded "hoophouse" where manure and bedding composts all winter in a dry environment. This immobilizes the nitrogen into complex carbon compounds before being spread on fields. Nitrogen in this form will mineralize over the course of the season, delivering plant-available forms of the nutrient just when the crop needs it.

This farm's profitability proves that other choices exist and are financially viable. Its biological complexity is equal to if not greater than the technological complexity of a large-scale hog confinement. Most important, it provides a true comparison of the benefits and problems of large-scale hog confinements. In addition to standard measures of economic success and resiliency, we need measures of social acceptability, impact on the local economy, ecosystem effects, and costs of enforcement and monitoring. Water relations, soil health, nutrient fluxes, pathogen spread, and heavy metal accumulation should be monitored and simulated at several scales from the farm to the regional watershed.

Finally, we should seriously consider and attempt to quantify the opportunity costs of converting to an economy in which hogs are raised entirely in corporate-owned, large-scale confinement buildings and all crops are grown on specialized cash grain farms. This cost is likely to be large in terms of reduced options for the future. If large-scale hog confinements are allowed to take over and the independent family farm infrastructure for hog production is squandered, the *option* for local nutrient cycling will be lost. This is a vast, uncontrolled experiment whose results we cannot easily anticipate until it is too late to act.

ACKNOWLEDGMENTS

The author wishes to thank D. Jackson, R. L. Huffman, D. Keeney, L. E. Lanyon, N. Lynch, D. D. Schulte, and G. R. Hallberg, and farmers T. L. Frantzen and M. M. Natvig for their contributions to this chapter.

REFERENCES

Antoun, H., S. A. Visser, M. P. Cescas, and P. Joyal. 1985. Effects of liquid hog manure application rates on silage corn yield. *Can J. Plant Sci.* 65: 63–70.

Baker, J. L., and H. P. Johnson. 1981. Nitrate—nitrogen in the tile drainage as affected by fertilization. *Journal of Environmental Quality* 10: 519–522.

Berendse, F., R. Aerts, and R. Bobbink. 1993. "Atmospheric Nitrogen Deposition and Its Impact on Terrestrial Ecosystems." In *Landscape Ecology of a Stressed Environment,* ed. C. C. Vos and P. Opdam, 104–121. London: Chapman and Hall.

Bitton, G., and R. W. Harvey. 1992. "Transport of Pathogens through Soils and Aquifers." In *Environmental Microbiology,* ed. R. Mitchell, 103–124. New York: Wiley-Liss.

Chang, C., W. R. Olmsgtead, J. B. Hohanson, and G. Yamashita. 1974. The sealing mechanism of waste water ponds. *Journal of Water Pollution Control Federation* 46: 1715–1721.

Davis, S., W. Fairbank, and H. Weisheit. 1973. Dairy waste ponds effectively self-sealing. *Transactions of the American Association of Agricultural Engineers* 16: 69–71.

Doppelt, B., M. Scurlock, C. Fressell, and J. Karr, 1993. *Entering the watershed: A new approach to save America's river ecosystems.* Washington, D.C.: Island Press.

Faulkner, W. 1959. *Light in August.* New York: Random House.

Hallberg, G. R., and D. R. Keeney. 1993. "Nitrate." In *Regional Groundwater Quality,* ed. W. M. Alley. New York: Van Nostrand Reinhold.

Hallberg, G. R., R. D. Libra, E. A. Bettis III, and B. E. Hoyer. 1984. *Hydrogeology, water quality and land management in Iowa.* Iowa City, Iowa: Iowa Department of Natural Resources, Geological Survey Bureau, Technical Information Series 22.

Huffman, R. L., and P. W. Westerman. 1995. Estimated seepage losses from established swine waste lagoons in the lower coastal plain of North Carolina. *Transactions of the ASAE* 2: 449–453.

Hurst, C. J. 1995. Removal of microorganisms during wastewater treatment. American Society for Microbiology General Meeting, Washington D.C.

Inouye, R. S., and D. Tilman. 1995. Convergence and divergence of old-field vegetation after 11 years of nitrogen addition. *Ecology* 76: 872–1887.

Iowa Department of Natural Resources. 1994. *Water Quality in Iowa During 1992 and 1993.* Des Moines: Environmental Protection Division, Water Resources Section.

———. Unpublished memo reporting conditions at fourteen swine waste lagoons operated by A. J. DeCoster, 26 September 1996.

Jackson, L. L. "Agricultural Industrialization and the Loss of Biodiversity." In *Protection of Global Biodiversity: Converging Interdisciplinary Strategies,* ed. L. Guruswamy and J. McNeely. Durham, N.C.: Duke University Press, in press.

Jackson, L. L., R. L. Huffman, L. E. Lanyon, D. D. Schulte, N. Lynch, D. Keeney, and G. R. Hallberg. 1996. "Water Quality." In *Understanding the Impacts of Large-Scale Swine Production: Proceedings from an Interdisciplinary Scientific Workshop,* ed. K. Thu. Iowa City: University of Iowa.

Jemison, J. M., and R. H. Fox. 1994. Nitrate leaching from nitrogen-fertilized and manured corn measured with zero-tension pan lysimeters. *Journal of Environmental Quality* 23: 337–343.

Kanwar, R., H. P. Johnson, and J. L. Baker. 1983. Comparison of simulated and measured nitrate losses in tile effluent. *Transactions of the ASAE* 26: 1451–1457.

Liebman, M., and E. Dyck. 1993. Crop rotation and intercropping strategies for weed management. *Ecological Applications* 3: 92–122.

Mapstone, B. D. 1995. Scalable decision rules for environmental impact studies: Effect size, Type I, and Type II errors. *Ecological Applications* 5: 401–410.

Marrs, R. H. 1993. Soil fertility and nature conservation in Europe: Theoretical considerations and practical management solutions. *Advances in Ecological Research* 24: 241–300.

McCurdy, M., and K. McSweeney. 1993. The origin and identification of macropores in an earthen-lined dairy manure storage basin. *Journal of Environmental Quality* 22: 148–154.

Meyer, M. 1996. How common is methemoglobinemia from nitrate contaminated wells? *Water Conditioning and Purification,* March: 78–80.

Midwest Plan Service. 1985. *Livestock Waste Facilities Handbook, 2nd Edition.* Ames Midwest Plan Service, Iowa State University.

National Research Council. 1993. *Soil and water quality: An agenda for agriculture.* Committee on long-range soil and water conservation, Board on Agriculture. Washington D.C.: National Academy Press.

Pennsylvania State University. 1994. *The agronomy guide 1995–1996.* University Park: College of Agricultural Sciences, Pennsylvania State University.

Ritter, W. F., and A. E. M. Chirnside, 1984. Impact of land use on groundwater quality in southern Delaware. *Ground Water* 22: 38–47.

Roth, G. W. and R. H. Fox. 1990. Soil nitrate accumulations following corn fertilized at various nitrogenrates in Pennsylvania. *Journal of Environmental Quality* 19: 243–248.

Schwab, G. O., D. D. Fangmeier, W. J. Elliot, and R. K. Frevert. 1993. *Soil and Water Conservation Engineering,* 4th ed. New York: J. W. Wiley and Sons.

Sparks, R. E. 1995. Need for ecosystem management of large rivers and their floodplains. *Bioscience* 45: 168–182.

Strange, Marty. 1989. *Family farming: A new economic vision.* Lincoln: University of Nebraska Press.

Sutton, A. 1994. Proper animal manure utilization. *Journal of Soil and Water Conservation Nutrient Management,* Special Supplement: 65–70.

Thu, K., ed. 1996. *Understanding the Impacts of Large-Scale Swine Production: Proceedings from an Interdisciplinary Scientific Workshop.* Iowa City: University of Iowa.

Verhoeven, J. T., R. H. Kemmers, and W. Koerselman. 1993. "Nutrient Enrichment of Freshwater Wetlands." In *Landscape Ecology of a Stressed Environment,* ed. C. C. Vos and P. Opdam, 33–59. London: Chapman and Hall.

Vitousek, P. M. 1994. Beyond global warming: Ecology and global change. *Ecology* 75: 1861–1876.

Vos, C. C. and P. Opdam, eds. 1993. *Landscape ecology of a stressed environment.* London: Chapman and Hall.

Warner, R. E. 1994. Agricultural land use and grassland habitat in Illinois: Future shock for midwestern birds? *Conservation Biology* 8: 147–156.

Part III

Justice and Equity

The three chapters in this section highlight issues of justice and equity, or the lack of them, when control of a basic resource such as food becomes concentrated into the hands of a few. Through detailed cross-cultural research among societies of varying social, economic, and political complexity, anthropologists have come to understand that access to, and control over, valued resources in a society is a source of power. Food is clearly a valued and basic resource, and control over its production and distribution is a source of considerable political clout whether it concerns pigs among the tribes of Tsembaga in New Guinea or pork among corporations in the United States. The message in these chapters is clear: a democratic political system is undermined when a valued resource becomes concentrated in the hands of a few.

In his contribution, Blaine Nickles, a life-long farmer, describes growing up and farming in a rural community in north-central Iowa. He details the dramatic social, environmental, and economic changes that occurred as large-scale swine operations began to surround his farm and neighborhood. Nickles then describes his quest into state politics to preserve his rural community and the quality of life for his neighbors. From local public hearings to the governor's office, a disturbing pattern emerges in which the efforts of farmers and other rural citizens are thwarted by political powers seemingly beyond their control. Nickles provides a wake-up call that issues of large-scale swine production are not just about pigs, but more fundamentally about who is tending the political fields.

From a different vantage point, Robert Morgan, former U.S. senator and state attorney general in North Carolina, describes how a large-scale swine producer in a position of power in North Carolina's state legislature quietly changed state laws to benefit corporate producers before their consequences were fully realized. Among others, laws were passed that protected large-scale swine operations from local zoning control by ensuring that they were considered "farming" and not industrial factory operations. Senator Morgan also discusses the influence of the swine industry on citizens' ability to receive fair treatment in the courts in order to protect their homes, property, and quality of life.

John Morrison, the executive director of the Contract Poultry Growers' Association, projects the swine industry's future should it continue down the path already taken by the poultry industry. He points to the political and economic inequities that exist between the few corporations who dominate poultry production and the poultry growers who contract with them. Morrison traces a number of disconcerting parallels between the poultry industry's past and the swine industry's impending future. However, he reminds us that there is a history lesson that can be learned and applied if policy leaders and the public have the political will.

These contributions are stark reminders that our food production and distribution system is integrally connected to the country's political health. As the food system goes, so too goes the rest of society. Highly concentrated and integrated food systems are related to highly centralized political systems. Food is not like other commodities, such as computers, telephones, and automobiles; it is as indispensable as the air we breathe and the water we drink. No matter the technological complexity or economic prosperity of a people or a country, its foundation is rooted in the way it makes food.

Chapter 7

An Iowa Farmer's Personal and Political Experience with Factory Hog Facilities

Blaine Nickles

I am a sixty-seven-year old lifelong resident of Wright and Hamilton counties in north-central Iowa. I grew up on a family grain and livestock farm and started farming in 1950 with a grain and livestock operation of my own. My wife and I had a beef cow herd during the early years, then started buying calves and yearlings to feed out for market. We bred and marketed hogs until recent years when we stopped farrowing and went to buying feeder pigs to feed out for market. We stopped raising cattle in the late 1980s and quit marketing pigs altogether in 1993.

After high school, I spent thirteen months in Korea with a "special units battalion" attached to the thirty first Infantry Regiment of the seventh Infantry Division. During that time we observed Korean farmers hauling human waste from the city of Seoul in wooden boxes mounted on high steel wheeled wagons pulled by oxen or horse. We referred to these wagons as "honey wagons." Korean farmers stored the waste in board-lined holes in the ground on their farms. Later they used it on their fields for fertilizer. Little did I know that forty-seven years later I would become involved in a controversy over how animal waste is stored and applied to the land in Iowa.

GROWING UP IN WRIGHT COUNTY

Life growing up on the farm in Wright County during the 1930s and 1940s was rather simple, self-sustaining, and hard work for farmers. Children attended one of two one-room country schools which were located two miles from each other and that had no electricity, were lighted with kerosene lamps, and had outside toilet facilities with no running water. In the 1930s, there was a family living on most 160 acre farms. There was the potential to support twelve to sixteen families per four square miles in the school district. Children attending the schools were part of a single class of first through eighth graders taught by one teacher. After the eighth grade, students could pay tuition and attend the high school in the town of their choice. Most public gatherings were held at the school houses. During the school year, the students took part in spelling bees and speech contests and, with the guidance of the teacher, participated in programs with skits and recitations for the benefit of the parents. They also had what they called "box socials," where each woman prepared a lunch and the boxes were auctioned off to the highest bidder, who got to share the lunch with the woman who prepared it. The teacher's box lunch usually sold for the highest price. At the end of the school year, there was an eighth-grade graduation and township picnic combination held at the Big Wall Lake Park where the families of Wall Lake Township took part in softball and other games, graduation ceremonies, and a big picnic.

I attended Sunday school and church in a small country church, Mulberry Center. We attended a Bible school class for one week during the summer at Wall Lake Center Schoolhouse taught by a farmwife, Nessa Bubeck. She was a big influence on my life and many others. The old Mulberry Center Church was moved to Webster City as a historical building. The Ed Miller family of rural Hamilton County was very instrumental in the development and preservation of that church and the farmland in that area. Most farm families attend church in town now, as only a few country churches are left.

The togetherness and relationships of social events spilled over to farm neighbors. Since farming in those years was more labor intensive, neighbors worked together to harvest crops, raise livestock, and butcher animals for their families' meat. Farm families also usually grew large vegetable gardens. Most farms had milk cows, hogs, and chickens, so they purchased very few groceries in town.

Our farm home did not have electricity until 1940 when the Rural Electrification Administration (REA) built transmission lines throughout rural areas. As technology progressed, farm mechanization moved very rapidly, except during World War II, when there was food and gas rationing. Also, very few farm tractors and machinery were built during that time since the government needed the steel products and food for the war effort. During the war years rural families became even more dependent on each other for working partnerships because many young men went into the armed forces to defend our country. Younger family members were called upon to do more of the farm chores because farm laborers were hard to find. But neighbors still found time to gather for card parties, homemade chili feasts, and homemade ice cream socials.

After World War II, the technology that had been developed to produce war materials was quickly put to use to produce tractors and other farm machinery. Farm work became more mechanized and less labor intensive. Farmers were able to farm larger acreages and produce more livestock with less man hours. In addition, younger people were able to obtain more education through junior colleges and universities and left the farm for better paying jobs in the cities. And as trade with other countries developed, the demand for farm products increased. In many cases, as older farmers retired, there was no family member to take over the operation, so the farm had to be sold or rented to a neighbor. As the size of farms and machinery increased, there were fewer neighbors working together. However, neighbors still help each other when illness hits, when there is a death in the family, or when other hardships strike. Neighbors still help with fieldwork, livestock chores, bringing in food (and lots of it), and help out through whatever crisis may arise. Many farmers still go to the local elevator or to a local restaurant for coffee in the morning to catch up on the "local news."

Lincoln Township in Wright County was developed and preserved for farming by families such as the McCormicks, Marshalls, Poncins, Woodleys, and Mechems, to mention a few. Many farmers such as myself have raised our families, produced grain and livestock, and worked with neighbors cutting silage, baling hay, and picking corn. While working in the fields, we would occasionally stop and visit with a neighbor across the fence in adjoining fields. We helped neighbors line up machinery for an auction when they were ready to retire. Since there are fewer farmers and larger tractors so that work gets done quicker now, we seldom meet at the ends of fields anymore. Farmers do not know as much about each other's operations as Randy Ziegenhorn describes in his chapter on farmer production networks.

During the 1930s, our township had about 140 farms. In 1995, there were only 61 occupied farm building sites. We have one dairy farmer left, six family pork producers, and four farmers with cattle. The rest are either grain farmers or are acreages occupied by people not farming. There are 19 acreages or farm sites occupied by nonfarmers. That leaves only 31 full-time grain farmers and 11 livestock producers in Lincoln Township.

I served on our Lincoln Township Agricultural Stabilization Conservation Service Committee for thirty years and did premeasurement and compliance measuring for the farm program from 1960 until the early 1970s. The farm program of the 1960s required diverting a percentage of a farm's corn acreage to a soil conserving cover crop. In complying with these requirements, the farmer received direct cash payments and could get price support loans for his corn. It was important that acreage measurements were correct. I also served eighteen years on the Board of Directors of the Farmers Cooperative Elevator in Dows, Iowa. During my time on the board, our cooperative placed pigs on member farms for them to feed on contract, sharing the profit 80 percent for the farmer-feeder and 20 percent for the cooperative. During the agricultural financial disaster of the 1980s, this was a way for some members to stay in the livestock business and make some extra money. Our cooperative also helped a young farm couple with a small land base expand into a three hundred sow farrowing facility through a program sponsored by Farmland Industries Cooperative of Kansas City.

Lincoln Township now has the potential (when construction is completed) to have four layer hen farms with 6.4 million layer hens, six 3,800-sow farrowing operations, and three 14,000 pig nursery operations, along with one 26,000-head hog finishing operation, one cement plant, and one feed mill. Since the 1980s, this township has certainly developed into a different community from the one of the 1930s or even the 1970s. What will it look like when all this construction is complete? How many family farmers and acreage owners will stay in Lincoln Township? It is likely that folks living on acreages and older family farmers will leave because of the odor and potential danger to their water supplies. As we see the rapid spread and concentration of factory-style livestock and poultry production corporations in Wright County, we fear for the environment. Because of this, our quality of life will be harder to maintain and the plight and disappearance of the independent family farmer and rural residents will be accelerated.

As a farmer and livestock producer, I certainly am not opposed to hog production, especially on family farms where the owners of the livestock live on, operate, and manage the farms. I see firsthand that what John Ikerd points out in his chapter is correct, that our rural communities are more viable and active where livestock and poultry are produced on family farms because farmers spend the money they earn in local communities. Family farms are different from large corporate livestock operations being built in several areas of Iowa and other regions of the country. Much of the earnings from these operations leave the community and go to nonresident investors. The few inputs purchased locally are for the initial construction of the facilities. Even though we all realize how important the pork industry is to the business communities of Iowa, quality of life, health, community stability, and the environment are also important.

HOG FACTORIES COME TO WRIGHT COUNTY

Wright County is home of the most rapid expansion of large-scale swine production facilities in the state. By 1994, Wright County had more construction permits (thirty-four) for these facilities than any other of Iowa's ninty-nine counties except one. The Iowa Department of Natural Resources (IDNR) issued construction permits to one producer in Wright County and one in Hamilton County Township (bordering Wright County) for eighteen farrowing (reproduction) facilities of 3,800 sows each. These facilities, when completed, will have production capabilities of 1,368,000 pigs per year. The combined farrowing, nursery, and finishing pig and poultry production facilities will be concentrated within just ten townships. The estimated amount of land needed for disposing the manure from these facilities will be nearly 73,000 acres. This is based on nitrogen requirements to produce a corn crop of 160 bushels per acre, the pounds of nitrogen produced by one sow and her offspring per year, and the nitrogen value of hog manure stored in pits under buildings and earthen basins. Our ten townships encompass nearly 219,000 tillable acres. The number of acres required to dispose of manure from these facilities each year would be 34 percent of the total tillable acres of these townships. The limiting factor for applying manure according to state legislation is nitrogen. However, as Laura Jackson discusses in her chapter, there is concern that because application rates are based on nitrogen, the phosphorous from manure will build up in the soil and pose a threat to rivers, streams, and lakes.

When completed, the total investment for the eighteen farrowing facilities, nine pig nursery sites, twelve finishing sites, five layer hen sites, and three pullet grower sites could be $243,293,738. Companies building these facilities include A. J. DeCoster and Iowa Select Farms. A. J. DeCoster came to Iowa in 1987 to build poultry facilities for pullet grower and layer hen units. He came to Wright County around 1990 when he purchased several farms and applied to the Iowa Department of Natural Resources for permits to construct hog and poultry production facilities. Since Wright County has zoning ordinances, he also applied for building permits from the county zoning commission. He is now the largest builder of large-scale hog production units in the county. By early fall 1993, DeCoster was issued permits to construct eleven separate confined hog production facilities in a concentrated area of Wright County, in addition to two pullet grower farms and two very large egg layer hen farms.

A. J. DeCoster is a Maine-based businessman, reportedly worth $120 million, and the seventh largest U.S. egg producer controlling over half the nation's brown egg market (Allen 1988). Prior to coming to Iowa to engage in pork production, DeCoster compiled a rap sheet for environmental and labor violations in other states. These included a fourteen-count action by the Maine attorney general and the Department of Environmental Protection for air and water problems (Goodwin 1987), an OSHA investigation and a farmworker suit related to asbestos exposure in chicken houses (Allen 1988), and a federal suit brought under the Migrant Agricultural Workers Protection Act that DeCoster provided unfit housing for his workers while harassing, threatening, and illegally evicting residents (Goodwin 1990). He recently received the largest fine ever issued by the Occupational Safety and Health Administration—$3.805 million (Jones 1997). This is not the kind of neighbor we are used to in Wright County, Iowa.

Since first coming to Wright County in 1987, DeCoster has purchased over 2,700 acres, established 26 hog production sites, and built over 130 intensive production facilities in the area and a township in an adjoining county (Wright County Treasurer and Recorder). With all sites fully operational, nearly 1.25 million hogs can be produced each year. This far surpasses concentration of production in any other area of the state and is five times the volume of average pork production in Iowa counties.

Joining DeCoster in this intensive swine production frenzy is another hog production company, Iowa Select Farms. Iowa Select Farms is run by Jeff Hanson, former head of Modern Hog Concepts. Modern Hog Concepts is a construction company specializing in the building of intensive hog production facilities. Prior to the establishment Iowa Select Farms, Modern Hog Concepts was contracted by DeCoster to build intensive swine production facilities. In fact, DeCoster ended up buying Modern Hog Concepts from Jeff Hanson in order to have his own company that could construct swine production buildings and manufacture necessary supplies. In 1994, Jeff Hanson was among the top ten financial contributors to Iowa Governor Terry Branstad's reelection campaign.

A small group of Wright County farmers and citizens met with the Wright County Board of Supervisors in the fall of 1990 to oppose the construction of large-scale swine and poultry facilities in the county. The supervisors responded that it was economic development for the county and there was not much they could do to stop it. The supervisors also viewed it as a source of increased county agricultural tax revenues.

I became involved because my wife and I farm within three-quarters of a mile of a large layer facility and live directly to the north and west of a heavy concentration of corporate livestock and poultry operations. We read newspaper reports from Maine, where DeCoster was operating poultry facilities, indicating that DeCoster was not a friend to the community or environment. At about this time, the summer of 1992, we started hearing reports about the negative social, economic, and environmental consequences of large-scale pork production for communities in North Carolina where the heaviest concentration of corporate hog facilities was occurring (Thu and Durrenberger 1994a, 1994b). We started seeing firsthand that similar problems were occurring in our own communities.

By the summer of 1993, odor problems and an infestation of flies from the egg layer and hen facilities plagued us while we worked in our fields. Neighbors living near a twelve thousand-head hog facility reported heavy infestations of flies when the wind blew toward their homes, as well as watery eyes, nausea, headaches, mental stress, depression, and anger resulting from the odors. One woman said she could not plan any more family reunions on her farm for fear of the wind carrying odor and flies to her home. She can no longer hang clothes on the line to dry because they absorb the odor. A neighboring grandmother told me how her children used to play in the water in a drainage ditch near their home in the summer. However, her grandchildren will not experience this farm fun because of slime in the water caused by runoff from manure in nearby fields. Odor is especially a problem when manure is sprayed on fields by huge sprinkler-type irrigation guns. People often have to leave their homes for several nights when this occurs. One mother described how friends of her seven-year-old son will not come to the farm to visit or stay overnight because of the horrible smell.

As of the summer of 1995, a 3,400 sow farrowing unit and a 14,000 pig nursery unit, both with two large anaerobic manure lagoons, operate 1.5 miles southeast of our home. When the wind is from the southeast we have to close our windows and cannot be outside because of the nauseating odor. Occasionally we experience odor from a mega site four miles away which has 26,000 hogs and 1.4 million layer hens on the same operation. As expansion of these large facilities continues in our community, we personally see many instances of the social disruption that Laura DeLind describes in her chapter, as hard feelings grow between neighbors and even relatives over the sale of land to DeCoster or other large-scale operators, such as Iowa Select Farms.

We and other long-time residents are concerned with a number of problems like those described by Laura Jackson in her chapter in this book, including the contamination of lakes, rivers, and streams from manure applied to the ground surface without being incorporated into the soil, manure applied to frozen or snow-covered fields, and airborne application of manure. There is the possibility of groundwater contamination from nitrates and coliform bacteria caused by overapplication of manure to fields located near rivers, streams, lakes, and agricultural drainage wells. Manure can also leach from earthen basins and clay-lined lagoon-type manure storage facilities. Manure can leak through cracks in the foundations of holding pits under confinement production buildings or leach into drainage tile lines that have not been capped.

I like to tell my grandchildren of how when I was a youngster I had to walk to and from a one-room school I attended three-quarters of a mile away and it was up hill both ways. There was a small creek running under the bridge on the road, and in the pasture near the road was a tile outlet draining water from the soils from our farm into the creek. The water from the tile was clear and pure enough to drink from, and we did. No one dares drink from these field tile outlets today.

In Wright County, there are thirty-eight registered agricultural drainage wells. Water drains from fields into these wells, which are a direct route to our aquifers that carry water underground. These drainage wells are the only outlet for drainage from these fields. DeCoster has built or plans to build hog or chicken facilities on ten sites that have twelve drainage wells on them. One of these is a mega site with a 26,000 hog capacity and 1.4 million layer hens on an area with two drainage wells. If a leak occurs, it will be a direct route of contamination to our underground water aquifers. Waste storage and disposal on these farms is an environmental accident waiting to happen.

Along with water quality, we are concerned with the quantity of water being used by these large factory facilities. They use millions of gallons per day for watering hogs and chickens. How long will our water in underground aquifers last? We are also concerned about the prime cropland taken out of production for these facilities.

Part of the community impact occurs during the construction phase caused by construction trash and material being blown into road ditches and neighboring farm fields creating extra clean-up expenses. Heavy trucks are required to haul in construction materials, and when construction is completed, to haul in feed and animals, as well as to remove heavy loads of fattened animals and manure. This imposes a burden on the county's road fund. Some of the facilities are built on grade B roads, which are minimum maintenance because of the previous lack of traffic. Facility operators are now requesting these roads be upgraded to level A with better surfaces and more maintenance at added cost to the county.

While a university study from North Carolina uses a complex mathematical formula to show how large-scale facilities decrease neighboring property values (Palmquist, Roka, and Vukina 1995), our firsthand experience provides the real measure. In one instance, a young couple did extensive remodeling to their house on a six-acre tract. A large-scale swine production facility was built a half mile away. They decided to sell their acreage and move because of the odor. The only people interested in buying was a young couple, one of whom, the husband, worked in a large hog confinement facility. They purchased the acreage at a price considerably under what the sellers had invested in it. There are instances where after DeCoster buys a property next to them, a neighbor will sell out to him because they do not want to live that close to a hog or poultry factory. This occurs even though they had no intention of selling prior to their neighbor selling out. For example, DeCoster purchased a parcel of land to build a 1.8 million layer hen facility. An elderly couple owned adjacent property, with their home situated near the fenceline near the proposed building site. They were more or less forced to sell because of the fear of odor and depleted or contaminated water in their wells.

Additional tax revenues expected by some county supervisors have not materialized. Under the Iowa Industrial New Jobs Training Act, operators of the new hog buildings have been eligible for Tax Increment Financing Funds through a local community college. Under the act, community colleges were able to enter into an agreement with an entering industry, in this case a corporate hog facility, to sell bonds to help finance "on the job training." Property taxes were then funneled to the community college and the owner to pay for the training of workers in the hog facility and pay off the principal and interest of certificates used to help establish the hog company. Our county gets to keep only 25 percent of the tax money generated by these hog factories, while the rest goes to the local community college and the owner to train the employees who work in them. The agreement has been changed recently, however, so the counties are allowed to retain a larger share of the new taxes generated.

GRASSROOTS RESPONSE

After learning of the influx and concentration of large production facilities during the fall and winter of 1993, I became involved with other farmers and citizens of Wright and surrounding counties. We held several small, preliminary meetings and decided to form a group to see what we could do locally and legislatively to solve some of the problems we saw in Iowa and other parts of the country where heavy concentrations of factory hog and poultry facilities were emerging. Early in February 1994 we met and named our group, the Organization for the Protection of the Environment (OPE). Arden Tweeten, a farmer from the Woolstock area was named chairman, and a treasurer and secretary were appointed as well. I have been part of this organization and a related Executive Committee that met with various farm commodity groups. We held an informational meeting February 15 in Woolstock with several legislators and local supervisors in attendance. We then held a statewide information meeting in the Clarion High School gymnasium, attended by nearly one thousand people, most of whom were farmers concerned with large livestock and poultry confinement issues.

OPE met with a variety of state representatives and farm group leaders. Throughout February and March 1994 we met with experts from Iowa State University College of Agriculture and their Extension Service who talked about manure management, groundwater protection, and drainage wells. We also met with state legislators, State Farm Bureau Board members, and representatives from the Iowa Pork Producers Association. We met with candidates for governor since it was an election year—congressional Representative Fred Grandy and State Attorney General Bonnie Campbell—and we met with Governor Terry Branstad on March 22, 1994.

Governor Branstad made it clear from the beginning that he was in favor of large-scale pork production in Iowa and, along with the Iowa Pork Producers Association and Iowa Farm Bureau, was concerned with keeping Iowa number one in U.S. pork production as measured by the total number of hogs produced. He informed us he was going to appoint an environmental agricultural task force to respond to our concerns by studying the environmental impacts of large-scale livestock and poultry feeding facilities. In his press release, Branstad announced the official purpose of the task force:

> I've assembled this committee to gather facts related to several environmental issues of livestock production, including nutrient management, facility location, design. This includes nutrient application to soils of the state. The committee will also serve as a sounding board for Iowans interested in these topics to present additional information or questions. After collecting good solid facts and public opinion, the members of this committee will develop recommendations on how the state should proceed in addressing these concerns (Office of the Governor, April 4, 1994).

The governor allowed one member of our OPE group to be on the committee. I was appointed and served along with two Iowa State senators, two members of the Iowa House of Representatives, two representatives from the Iowa Cattleman's Association, two from the Iowa Pork Producers Association, two representatives from the Iowa Poultry Association, one representative from the Iowa Sheep Industry Association, one member of the Iowa Dairy Products Association, two persons from the banking industry, one from the Des Moines Chamber of Commerce, one from the Iowa Farm Bureau, one from the Leopold Center for Sustainable Agriculture, one from the Izaak Walton League, one from the Iowa Department of Natural Resources, and one from the governor's existing Livestock Revitalization Task Force. The Dean of the College of Agriculture at Iowa State University, David Topel, served as the chairman. The unbalanced makeup of the committee, with heavy industry and business representation, made the possibility of developing reasonable policy recommendations to deal with large-scale livestock operations out of the question. It was like asking a fox to guard the chicken house.

The Republican Governor's Task Force was established as part of the governor's existing Livestock Revitalization Task Force headed by Robert Furleigh, former head of Iowa's state Agriculture Soil and Conservation Service office. Bypassing the Democratic state secretary of agriculture, the Livestock Revitalization Task Force was formed in March 1993 with its chairman assuming an office next door to that of Governor Branstad in the state capitol. This task force was charged with increasing Iowa's 1 percent share of the national livestock market to 5 percent by the turn of the century. As part of this effort, the governor approved a $3 million agricultural research fund for Iowa State University's College of Agriculture, headed by Dean David Topel, his other task force leader (Muhm 1993).

The head of the Governor's Task Force investigating swine industry problems was beholden to the governor for research funds. One of the primary builders of large-scale swine production facilities in our area was a major contributor to the governor's campaign. These relationships made it difficult for us to get our problems fairly addressed by the legislature. PrairieFire, a rural farm advocacy group, and the Iowa Farmers Union helped us lobby the 1994 Iowa legislature to pass legislation to protect our rural environment introduced by Representative Stewart Iverson. Despite these efforts, manure management legislation did not pass the 1994 session. Our concerns and efforts fell on deaf ears.

While our efforts failed, the Governor's Task Force continued in the face of increasing negative publicity concerning corporate hog facilities in the midst of an election year. The task force met on April 12 and planned five public hearings across Iowa. In addition, tours of livestock facilities were conducted and meetings were held at the Iowa Pork Producers Association and the Iowa Institute of Cooperatives. Hundreds of citizens from all over the state testified, and oral and written testimony from experts was provided. Following the hearings, a writing committee was appointed to draft a task force report. Dean Topel assumed the chair of this committee, which included Harold Trask of the Iowa Pork Producers Association, Gene Troyer from the Iowa Poultry Association, Wythe Willey from the Iowa Cattlemen's Association, State Representative Delores Mertz, Pauline Novotny from Izaak Walton, and myself. We met in early October, and each member was charged to write suggestions for legislation. When the Pork Producers Association representative and I visited and discussed the issues and made plans for legislative suggestions, I was under the impression that the input of citizens who testified at the public hearings would be given consideration and that he would do more to help protect the safety of our agricultural drainage wells and quality of life. I took this responsibility seriously and reviewed hundreds of pieces of testimony provided during the public hearings, as well as written testimony from experts. I integrated this testimony into my draft of legislative recommendations.

When the full committee reconvened on October 26, two separate proposals were presented: one developed by the Iowa Pork Producer representative (Plan 1), and one I developed (Plan 2). When Dean Topel took the two plans from the writing committee meetings, he put them both in a proposal to the full committee, but he listed the Pork Producers proposal as plan 1 and mine as plan 2. Plan 1 was discussed first in the full committee and given a more favorable reception than any parts of plan 2. As a matter of fact, the committee never discussed plan 2 in full. I could not give the final draft my "yes" vote because it left out many important items which I thought should have been proposed to the Revitalization Task Force and ultimately the legislature. I voted "no" on the final committee draft. Dean Topel eventually adopted plan 1, which was primarily a reprint of the Pork Producers plan. The proposal was approved by a majority vote of the committee. My plan was not considered. The main differences between my suggestions and the proposal submitted to the full committee for consideration were as follows:

1. I was asking for a surety bond to be posted based on the size of the manure lagoon or earthen basins. The Pork Producers' plan called for an indemnity fund, a request weakened further by the state legislature.

2. My plan would have banned or restricted the use of irrigation guns for airborne manure disposal.
3. My plan called for more protection of our agricultural drainage wells, rivers, streams, and lakes by creating larger separation distances from livestock facilities and when spreading manure on frozen and snow covered soil.
4. Site selection and separation distances from residences and confinement buildings and manure storage facilities were greater in my proposal. Mine was based on a formula of two pigs per foot, starting with 2,500 butcher hogs and 1,250 feet and on up to 20,000 hogs which would be required to be 10,000 feet away from a neighboring residence.
5. They also would not have been allowed to build on environmentally sensitive areas, e.g., agricultural drainage-well farms or flood plains of rivers, lakes, and streams.
6. My proposal called for a combination of state regulations and local zoning authority.
7. My proposal gave specific suggestions for odor control.

Only weaker siting rules, which called for minimal setback distances, a manure management plan requirement, and an indemnity fund (later weakened in the state legislature), were included in the Governor's Task Force report. The vast majority of problems experienced by the citizens of Wright County went largely ignored. I read every bit of public and written testimony, but the report finalized by Dean Topel incorporated none of it.

I was not satisfied with the process by which the task force developed its recommendations or the actual written proposal, though minor parts of it were good. It did not go far enough to protect our environment and the people of Iowa. While the Task Force approved the final report for presentation to the standing Livestock Revitalization Task Force, the governor, and finally the state legislature, permits continued to be approved for fifteen more factory hog facilities and two layer hen facilities in Wright County, as well as three more hog facilities in a neighboring county. I believed that the representatives and executives from the Iowa Pork Producers Association, Farm Bureau, Iowa Cattleman's Association, and Iowa Poultry were the primary influence in writing and deciding the final draft of the committee recommendation.

In addition to the Governor's Task Force, a legislative interim study committee, a citizens task force made up of PrairieFire, the Iowa Farmers Union, the Iowa Citizens Action Network, the Iowa Farm Unity Coalition, and the Organization for Protection of the Environment was created to make concrete legislative proposals. The proposals made by the Citizens Task Force were similar to my own in that they recommended larger separation distances based on numbers of pigs per site. They also recommended posting bonds for total clean-up of sites, not just manure clean-up. They also recommended some local control for site selection and local manure disposal requirements; no use of the irrigation gun; liners for manure lagoons other than clay; and use of some method of odor control.

On February 21, the state House of Representatives Agricultural Committee held a public hearing. About one hundred citizens testified, some with specific legislative suggestions and almost all opposed to factory type livestock operations and in favor of independent family farms. Several lobby days were held at the state capitol by citizens from all across Iowa from the opening of the legislature until the livestock bill was debated. The bill that was deliberated and eventually passed (House File 519), generally followed the Governor's Task Force report. Our grassroots OPE group, the Citizens' Task Force, and many other citizens of Iowa were disappointed with the legislation. The sections of House File 519 that this author found most disappointed in are:

- The indemnity fund provision is not a continuous fund, it is only a one-time fee, and it should be a larger fee per animal unit.
- There should be larger separation distances between residences and livestock buildings and lagoons. There should be larger separation distances between agricultural drainage wells and rivers, streams, lakes, and drainage ditches.
- There should be a more specific provision for local control over site selection based on concentration and environmentally sensitive areas.
- Lack of recommendations for odor control. I think they should be required to put covers on and plant trees around all outside formed manure storage, lagoons, and earthen basins. In addition, chemical additives should be used in the manure storage facilities.

One of the most disappointing sections of the legislation is a clause providing lawsuit protection for livestock and poultry producers who get permits from the Department of Natural Resources. This makes it almost impossible to bring a suit against a permit holder. Our efforts to get legislation to protect our rural areas from corporate hog and poultry facilities were turned on their heads by a system that ended up protecting hog factories and their operators from us.

DISCUSSION

In the face of all this political wrangling and maneuvering, my community continues to be swamped by large-scale corporate livestock facilities. Thirty-nine corporate hog production facilities have been or are slated to be constructed within a small area of Wright and an adjoining county that can produce over 1.5 million hogs each year. This is three times the number of hogs produced by our two counties only three years ago (USDA 1992). In addition, layer facilities with nearly 6.4 million hens will be concentrated in the same area. We are experiencing six kinds of problems associated with this kind of concentration:

1. inadequate land to dispose of animal wastes,
2. odor,
3. potential water contamination,
4. displacement of independent family farmers,
5. disruption of community relationships,
6. lack of responsive and effective state policy.

In addition to these, market access for some independent family farmers is becoming a problem because of the large numbers of hogs being delivered each day to packing plants by large factories and contract feeders. The large packing plants prefer regular deliveries of large loads of hogs over small shipments from independent producers. The large producers seem to have a price advantage as well.

We express these concerns based on our personal experience of what is occurring in our community. Yet politicians generally have ignored our warning signals, and the political system that is suppose to correct these situations has not been responsive. As a result, several environmental disasters making local and national headlines have occurred in Iowa and other parts of the country. Just south of Wright, near the town of Blairsburg, 1.5 million gallons of manure leaked from a holding lagoon into an abandoned drainage tile. From the tile, the manure made its way into the Iowa River where it polluted a thirty-mile stretch, killing 8,000 fish according to the Iowa Department of Natural Resources. Another spill in northeast Iowa resulted from manure leaking into an underground drainage tile and then into the Little Volga River, killing 17,000 fish. These spills occurred in the wake of an even greater disaster in North Carolina reported by CNN and the *New York Times* where 22 million gallons of raw hog sewage from a hog factory leaked into the New River. These disasters, combined with two additional spills in Iowa since, show that our concerns are valid. It apparently takes a disaster to get politicians to start acting, yet these and other problems can be avoided if policy makers and state leaders would simply listen to what we are saying.

Many of us are concerned that thousands of acres of the most productive land in the world here in the nation's heartland are being covered with livestock and poultry facilities and will never again grow grain crops. Sacrificing air and water quality for factory hog and poultry production in Iowa is not worth it. It is possible to encourage responsible livestock and poultry production without environmental and community sacrifices, but it is necessary to have better state policies to accomplish this.

In rural Iowa, neighbors are important because we help each other out in time of need, we invest our money and hearts in friends and family, and we are proud to work hard to leave something for our children to inherit. This is the foundation of our country. We now have corporate neighbors who think first about their profits, care little about the needs of the community, and are leaving a trail of waste for our grandchildren to inherit. If this kind of thing is happening in rural Iowa, then it must also be happening elsewhere. We had better wake up, smell the coffee, roll up our sleeves, and start cultivating the political fields ourselves. There are a lot of weeds there, and no one else is going to do the work for us.

REFERENCES

Allen, Scott, "The DeCoster Mess," *Maine Times,* 12 August 1988.

Goodwin, Jody, "DeCoster Cited in 14-Count State Action," *Lewiston Daily Sun,* 20 August 1987.

———. "Workers File Suit against DeCoster," *Lewiston Sun Journal,* 5 April 1990.

Goudy, Willis, and Sandra C. Burke. 1995. *Iowa's counties: Selected population trends, vital statistics, and socioeconomic data.* Ames: Department of Sociology, Iowa State University.

Jones, Heather C., "DeCoster, OSHA Reach Settlement," *Feedstuffs,* 26 May 1997, p. 5.

Muhm, Don, "Branstad to Bolster Livestock," *Des Moines Register,* 4 March 1993, Sports and Business section, 1 Star edition, 8.

Palmquist, Raymond, Fritz M. Roka, and Tomislav Vukina. 1995. *The effect of environmental impacts of swine operations on surrounding residential property values.* Department of Economics and Department of Agricultural and Resource Economics, North Carolina State University.

Thu, Kendall M., and E. Paul Durrenberger. 1994a. North Carolina's hog industry: The rest of the story. *Culture and Agriculture* 49: 20–23.

———. 1994b. Industrial agricultural development: An anthropological review of Iowa's swine industry. Paper presented at Twentieth Annual Meeting of the National Association of Rural Mental Health. July 1–4, Des Moines, Iowa.

United States Department of Commerce. 1992. *Census of Agriculture.* Washington, D.C.

Chapter 8

Legal and Political Injustices of Industrial Swine Production in North Carolina

Robert Morgan

North Carolina is a large state, extending about 550 miles from the Atlantic Ocean west to the Tennessee border and about 250 miles from the Virginia line south to Georgia and South Carolina. It is made up of three distinct regions: the mountainous west, the central piedmont, and the eastern coastal plains.

The eastern region has been described as a "sleeping giant." Predominantly rural with only scattered manufacturing, it has the opportunity to develop in a planned, environmentally sound way. Traditionally, this has been an area of relatively small farms which grew tobacco, corn, cotton, wheat, and soybeans. Those farmers who raised livestock usually produced enough corn and other feed grains to provide their own feed. It is perceived as being unspoiled by industrial pollution. However, it is now apparent that its rivers are becoming clogged and spoiled by the sewage of piedmont cities, industrial pollution, and agricultural runoff.

In 1984, there were more than twenty thousand hog farms in North Carolina populated by 2 million hogs. More than a fourth of these farms, mostly in the east, had fewer than five hundred hogs, and only a small fraction had more than two thousand. A decade later, the number of hogs catapulted to over 7 million, but the number of hog farms fell below five thousand. In little more than a decade, the number of hogs produced nearly quadrupled, while the number of farms dropped by three-fourths. Hog farming suddenly became big business as only 2.5 percent of these farms had fewer than five hundred hogs, while 84 percent had over two thousand each.

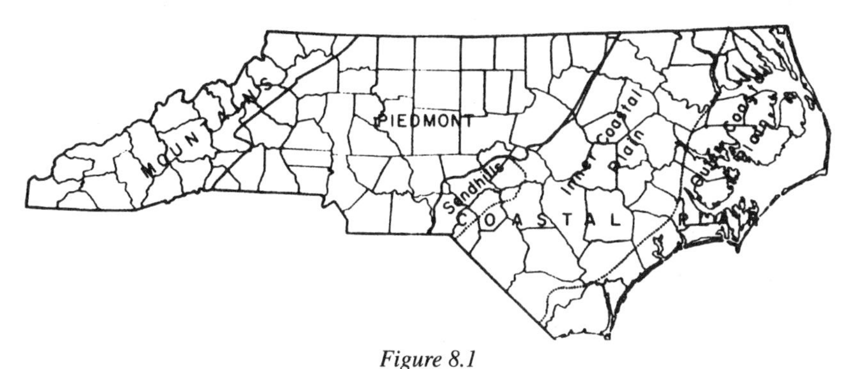

Figure 8.1
Major Terrain Regions of North Carolina
Source: Atlas of North Carolina, Richard E. Donsdale, Director and Chief Cartographer. The University of North Carolina Press: Chapel Hill, 1967.

QUIET POLICY CHANGES BENEFIT CORPORATE HOG PRODUCTION

Along with the size of operations, the pattern of ownership changed, and the independent owner-operated hog farm is becoming a thing of the past. Today, most of the farmers who raise hogs on their land do not own those hogs, but produce and raise them on a contract basis for fees set by a handful of corporations that dominate the industry. The consequences of contract production that John Morrison describes in his chapter are becoming evident in North Carolina. Although farmers take on huge debts to build the necessary facilities to raise hogs, their contracts with the corporations typically extend for only a year at a time and can be canceled with only three months' notice.

While these changes were taking place on the farm, little-noticed changes were also being made in the state legislature. These changes were spearheaded by Senator Wendell Murphy, a highly respected legislator in the late 1970s and 1980s, and the nation's largest hog producer. People knew that Senator Murphy was bringing about innovations in hog production in his home county and most approved because tobacco was on the way out and expanded hog farming could help replace it. What we did not know was that state laws were quietly being amended to give the corporate hog industry virtually free reign.

No one paid much attention when the legislature passed a "right to farm" law in 1979; the family farm has always been the bedrock of American society and no one argued with legislation to protect it. But this law provided that no agricultural operation shall become a private or public nuisance by any changed conditions in or about the locality after the operation has been in operation for more than a year, except where the nuisance is caused by negligent operations.

A similar change has occurred recently in Iowa law. Another significant piece of legislation restricted counties' zoning authority. Counties had never been allowed to zone a "bona fide" farm, but a bill enacted in 1991 defined a "farm" to include all types of livestock and poultry production, even though many such operations are designated as intensive animal feeding operations by the Soil Conservation Service.

These changes meant that protections intended for the family farmer were extended to corporate hog farms. But the term *hog farm* is a misnomer; these new facilities are really hog factories, with none of the attributes of traditional farming operations. The animals spend their lives in specially designed intensive growing facilities where they are fed and watered through automated distribution systems. Although they are never in contact with the farmland and are not visible to passers-by, their urine and feces are. These wastes are periodically flushed to open manure lagoons then sprayed on open fields.

Not only did the legislature protect the livestock and poultry industry from regulation or law suits, but also it awarded them a series of tax breaks, including exemption from sales tax on all materials and equipment used for livestock or poultry facilities. The local property tax on feed used in livestock production was eliminated, and corporate hog farms were exempted from paying a state inspection fee on ingredients used to feed their animals. The hog industry's growing political power was due not only to the efforts of individual legislators such as Senator Murphy, but to large and continuing contributions to candidates; it takes a great deal of money to get elected these days, and the industry was generous in its support of officials who might be able to help it in the future.

BATTLING AN ENTRENCHED SYSTEM

This situation was brought to my attention in early 1992 when a group of homeowners and small farmers asked me to represent them in their efforts to protect themselves against damage done by a large-scale hog operation. Earl and James Lee bought a tract of less than one hundred acres in the middle of Meadow, a pleasant community with a couple of churches, a school, and some modest homes, many of which had been in the same family for generations. The land was sold without anyone knowing what was intended, and, almost overnight, three intensive swine facilities which held three thousand hogs total were constructed. The new owners also constructed what they called a "lagoon."

It was not a lagoon. That word conjures up images of a tropical pool with a beautiful woman swimming in it. This was a cesspool. These facilities were within a few hundred feet of several homes and within 1,500 feet of a subdivision that was being developed. A field on which waste was sprayed abuts the highway and several more homes. Later, this operation would expand to ten swine facilities, which held ten thousand hogs.

When the people of Meadow learned that their new neighbor was to be a hog factory, they complained to the county board of commissioners, but they were met with little sympathy and no relief. At that time, there was little awareness of the potential environmental damage these operations could create. Factory farms were still thought of as an asset to the county, and at least two of the county commissioners had large operations themselves. Even if the county commissioners had been inclined to adopt zoning regulations to protect the community, they were prohibited from doing so by the 1991 legislation. Theoretically, county health officials could still declare the hog operation a nuisance, but the industry threatened the county with lawsuits that would have been devastating for a poor rural county.

The local people were desperate, for they were assailed by odors from the facilities, from the lagoons where the wastes were stored, and from fields sprayed with manure. Since these odors tend to move in plumes, they can travel long distances with minimal loss of their potency and affect different places at unpredictable times. As if the stench from the lagoons was not bad enough, fields right beside some houses were periodically sprayed by irrigation guns with this foul waste. With the threat of intolerable stench always present, the swine facility's neighbors could never safely plan outdoor activities and were, in effect, denied the domestic tranquility guaranteed by our Constitution. The odors would descend without warning on cookouts and birthday parties and sometimes made routine chores like hanging out laundry or mowing grass impossible. One hog operation neighbor, who lived in a small house that had been in his family for generations, had a respiratory condition, and his family and doctor believe that the fumes contributed to his death.

Not only were the people of Meadow unable to obtain relief from the county board, but the influence of the hog industry was so great that they could not find any lawyer in the county who was willing to represent them. I had retired and opened offices in nearby Lillington as well as the capital, Raleigh, and agreed to take on the case, naively thinking that the matter simply involved a common law nuisance and we would just have to prove that the hog operation gave off such noxious, nauseating, and sickening odors as to unreasonably interfere with its neighbors' enjoyment of their property. The law is clear in North Carolina that a person who intentionally creates or maintains a private nuisance is liable for the resulting injury to others regardless of the degree of care or skill exercised by him or her to avoid such injury.

Before I became state attorney general of North Carolina, I spent some years as a country lawyer. We lawyers all knew each other, and we took our cases down to the courthouse where we went up against each other but all played by the same rules. When big business steps in, it no longer works that way. This case, and a nearby operation involving four hog facilities, turned into complex and very expensive lawsuits, with the defense funded largely by the North Carolina Pork Producers Association and the industry. As the association and the industry's magazine said, the industry saw this "as a test case and one that could result in a landmark decision detrimental to all pork producers."

The defendants brought in a law firm with more than fifty lawyers and assigned several to the case. The attorneys started taking depositions—written statements from parties who are examined under oath, which are admissible in evidence. I understood why they wanted to depose a few complainants, to see what they had to say. But these lawyers, who had virtually unlimited time and money, took between twenty and thirty depositions. This not only put pressure on the plaintiffs, who were unaccustomed to legal proceedings, but forced my small firm to expend an inordinate number of hours, although we were being paid only a small fee for the whole proceeding. One example of the extremes to which these lawyers went is that they obtained a court order to permit the defendants to go on the property of the landowners and take urine samples of their dogs and cats, contending that these pets were the source of the odors.

The industry's resources helped the defendants in more subtle ways. When attorneys for the plaintiff, armed with a court order, inspected the facilities, they found that the owners were accompanied by several professors from North Carolina State University, a land grant college with responsibilities for agricultural research. It is perhaps relevant that Senator Murphy not only serves on the university's Board of Trustees, but has contributed several hundred thousand dollars to its programs. These same professors had taken part in more than a dozen research projects funded by the Pork Producers association and one had served on its board of directors.

These ties explain why we could not get an expert anywhere in the state to testify for the plaintiffs and why these same four professors later testified that the facility's design met state-of-the-art standards and was not negligently operated. This testimony repudiated their previous writings, which stated that producers should build aerobic, rather than anaerobic, manure lagoons if noxious odors were of concern to their neighbors. But those would cost more, so the corporate hog producers ignored that advice.

We ran into the same kinds of problems when we tried to find real estate appraisers or salesmen who would testify about the value of property. They were so afraid of losing the hog industry's business that the complainants could not find a anyone willing to work with them.

The corporate hog producers argue that the contribution they are making to the rural economy outweighs the disadvantages. It is true that the rule of law is the measure of loss to one party weighed against the advantage to the other from granting or refusing injunctive relief and that the court will consider this. But a balancing of the equities does not favor the corporate farmer, for while the corporation may benefit from the operations, there are few benefits to the community. The sworn testimony of the owners of the three thousand-hog operation showed that it had no employees, as the operation was fully automated, and that the feed was brought in from the Midwest rather than purchased locally. The ten thousand-hog operation had only three part-time, minimum wage employees.

The Court of Appeals would later rule that the "right to farm" law, which restricted the statutory grounds for bringing a nuisance action, did not protect a so-called hog farmer from such actions if he had made a fundamental change in the way the land was used. But at the time we went to court, the law was still in effect, and we had to fall back on the common law remedy. The common law—the unwritten rules that have governed for centuries—basically says that you can use your property any way you want as long as you do not interfere with the other person's right to use his property.

At trial, we showed that the plaintiffs had lived in their homes long before the defendants began their operations and there had been no previous odor problems. The defendants already had a five hundred-acre farm with an intensive livestock operation and could have located their new operation there instead of inflicting this nuisance on Meadow. The defendants did not discuss their plans with anyone in the community before starting their operations. In fact, word went around that they were planning a horse farm. The plaintiffs complained as soon as the first facility was loaded with hogs, but the defendants, instead of responding, tried to mislead the plaintiffs by assuring them that the odors would go away. They then asked for thirteen more building permits, using minimum permissible design criteria for the facilities, and later they purchased forty-one acres of adjacent land upon which they planned to expand.

REFLECTIONS

We lost our case, but we helped sound an alarm about the effects of these swine factories. Public interest escalated when, in the summer of 1995, the dike of a supposedly "state-of-the-art" manure lagoon broke and 22 million gallons of hog waste poured across roads and crops and into the New River, causing massive fish kills and threatening the area's water supply. After this, no one could claim that these hog factories were not a potential environmental hazard. The state's leading newspaper, the *Raleigh News and Observer,* then came out with a hard-hitting series of articles on the subject, earning the paper a Pulitzer Prize.

Of course, the hog interests fought back, with intensified lobbying and a public relations campaign, trying to show that the industry was composed of small farmers. They got legislation passed requiring people to go to mediation before they could bring a nuisance action against hog producers. They defeated legislative efforts in 1993 to impose stricter requirements on hog farms and ensured that a commission created to study the problem was biased in their favor. The state did eventually tighten regulations on the placement of facilities, several suits are now pending against facilities, and various regulatory proposals are currently pending in the state legislature.

As a former state attorney general, I have worked with enforcing antitrust and restraint of trade laws and am increasingly concerned about a situation where 85 percent of hog production in the state is controlled by a few large corporations. There are no laws to protect the contract farmer, who typically takes on heavy debts to construct facilities, with no assurance that his contract will continue. The problems caused by these hog factories will not simply go away. In 1995, state inspectors checked about three thousand hog farms and cited one-third of them for waste management problems, finding that almost five hundred had lagoons that were in danger of overflowing. Protections designed to preserve the family farm have been perverted to protect these giant swine factories and are actually helping to drive the independent farmer out of business. But hog production can be carried on successfully and still be beneficial to the community if the industry will take the necessary steps to protect the environment and respect the interests of the contract farmers and their communities. If the industry fails to take such leadership, public pressure eventually will force changes; no group is powerful enough to go against the public interest indefinitely.

Chapter 9

The Poultry Industry: A View of the Swine Industry's Future?

John M. Morrison

Many in agriculture view recent developments in the swine industry as perilously similar to the history of the poultry industry. While Wall Street analysts perceive this to be a tremendous success, there are many problems stemming from changes in industry structure. In this chapter, I provide a critical overview of the poultry industry, including an outline of serious social and economic problems. I then draw a parallel linking these flaws to current developments in the swine industry.

DEVELOPMENT OF THE AMERICAN POULTRY INDUSTRY

As families struggled to feed and clothe themselves during the Great Depression, a number of entrepreneurial enterprises were spawned. One family in the midst of the Ozarks found they could grow enough vegetables to feed themselves and have some to sell. Often, their neighbors did not have hard cash for produce, so John Tyson, the patriarch of the family, took chickens in trade and marketed them in town. Observing the demand for chickens, he built relationships with many families and developed a supply base. As his chicken supplies grew and he began to expand his market area, he found he needed a better way to maintain the market condition of chickens on his longer trips. In the midthirties, he devised a means of attaching feed troughs and drinkers to his vehicle which permitted him to expand his marketing as far away as Chicago.

For several years, John continued to buy, transport, and sell exclusively chickens. He also began looking into the husbandry of poultry production. By 1947 he was raising most birds on his own and decided to incorporate as Tyson Feed and Hatchery. He focused on chicken production and sales into the early fifties, and in 1958 he built his first processing plant in Springdale, Arkansas.

There were many similar entrepreneurs scattered across the country, including Frank Perdue, Bo Pilgrim, and Cliff Lane. These men started in their backyards and built regional and national operations that contributed to the industrialization of the nation's poultry industry. Each managed the production of fertile eggs, hatching of chicks, milling of feeds, raising, slaughtering, processing, and marketing of the product within a single company. The concept of controlling basic raw materials, processing, and product marketing had been used a century earlier by the large industrial steel and auto enterprises, but this was the first major venture in agriculture. Vertical integration in the poultry industry thus became a basic business structure.

PRESENT DAY POULTRY INDUSTRY

The efficiencies of a vertically integrated production system brought cheap, readily available products to markets, and the demand for chickens grew. Early integrated companies experienced difficulty in expanding because of huge capital demands. Companies wanting to extend operations geographically needed capital to build processing plants, hatcheries, feed mills, and growout houses. They found company capital requirements were reduced by half when the growers or farmers put up the other half through contract arrangements. In building growout houses, the growers collectively end up as 50 percent partners.

Almost all broiler production in the United States today is done through contracts with independent growers. A few processors have a small number of their own production units, but company farms have not proven to be as cost efficient as independently owned farms. Why own the farm when you can own the farmer?

The typical broiler contract arrangement calls for the company to provide day-old chicks, feed, medication, and technical advice to raise the birds to market age, usually a six-to eight-week period. The farmer contributes fully equipped poultry houses built to company specifications, utilities, dead bird disposal, manure disposal, and labor to care for and raise the birds to market. When the birds are market age, the company picks them up, hauls them to the processing plant, weighs them, and pays the grower on a per pound or per head basis.

How do the parties fare under this contract arrangement? Tyson, the largest poultry company in the world, has enjoyed returns on equity in excess of 20 percent for a number of years (Atlas 1994). These record returns were only interrupted in 1994 when the company lost $200 million on an investment in Alaskan Fisheries. Even with this loss, their five-year average return on equity only dropped to 13 percent (Heuslein 1995). ConAgra, the second largest poultry processor, generated returns on equity averaging 18 percent over the past five years (Heuslein 1995: 174). Processors have done exceedingly well compared to average returns of 11 to 12 percent for industry in general.

The contract poultry farmer has not fared well during this period of processor prosperity. Studies of grower income (Clouse 1995) have recorded nominal net earnings of about $4,000 per year per poultry house. With an average farm of three grower houses, a $12,000 annual income provides a bare subsistence living for a farm family in the face of as much as $250,000 to $300,000 in poultry operation debt (Gallagher 1995). In several areas across the country, growers have not had an increase in contract pay during the past ten years, yet operating costs for supplies, gas, electricity, and equipment have more than doubled.

Growers also complain that companies constantly require upgrades in equipment for their production houses to maintain state-of-the-art facilities, yet no increase in contract payments is given to pay for equipment updates. Growers usually accede to company demands for fear that no more birds will be placed on their farm if they refuse.

Why do farmers enter into an indentured relationship with such little economic return? The answer can be understood by comparing growers who have been in business over fifteen years with those who recently built their poultry operations. When the older segment of growers began contracting, they were likely dealing with a processor who was a member of the community in a manner similar to what Blaine Nickles and Jim Braun describe in their chapters. The processor probably went to the same church as many of his contract growers, had children that went to school with his growers' children, and by virtue of these social ties had a real interest in the welfare of the producer. The grower was equally concerned about the economic well-being of the processor, and each prospered under this arrangement. As companies grew and were bought out by outside companies, this interpersonal relationship disappeared. The community's social covenant was displaced by purely profit driven contracts—the industrial paradigm John Ikerd discusses in his chapter.

Growers entering the industry within the last fifteen years were influenced by the following factors:

1. Increasingly, economic depression in rural areas and any option for saving the family farm is welcomed.
2. A farmer's major asset is land. A small farm in the fifty to two hundred acre range cannot sustain itself with cultivated crops, but with government guaranteed loans, the land can provide equity to finance major building and equipment costs to begin poultry production.

3. Rural society is characterized by a tradition of trust, making farmers vulnerable to incomplete information, false promises, and exploitation.

What caused the great disparity in economic returns and the tenuous relationship between the grower and the processor? To understand this relationship it is important to examine the major factor governing it—the contract. Included below (see table 9.1) is an example of the characteristic of a typical contract used by a major poultry processor. A comparison to other contracts reveals similar provisions and structure throughout the industry.

Tyson Foods has contract growers from Maryland on the eastern seaboard across the broiler belt to Texas. Tyson's standardized contracts used in all its poultry units provides a good example to examine.

Under this arrangement, Tyson furnishes chicks, feed, medication, and technical advice, while the grower provides housing, poultry equipment, utilities, labor, and waste disposal. This practice is common throughout the broiler industry. Once the relationship is established, Tyson brings day-old chicks to the grower's farm, places them in the care of the farmer, and provides feed and medications as needed. Tyson picks up the birds in the six to eight weeks, depending on the size of the birds needed for market, hauls the birds to the processing plant,

Table 9.1
Tyson Contract Responsibility and Accountability Summary

Contract Item	*Quality Accountability*	*Framework and Requirements*
Furnished by Tyson		
Contract language		No grower input, "*take it or leave it*"
Chicks	None	
Feed	None	
Medication	None	
Vaccines	None	
Management	None	
Technical advice	None	
Grower pay		Pay based on grower ranking system
Furnished by Grower		
Poultry house	Yes	Must meet Tyson specifications
Equipment	Yes	Must meet Tyson specifications
Utilities	Yes	Must meet Tyson specifications
Labor	Yes	Must fulfill Tyson management guidelines
Bird disposal	Yes	Must meet governmental regulation
Manure disposal	Yes	Must meet governmental regulation
Equipment update	Yes	Change/add systems on Tyson request

weighs the live birds, and usually processes them for a specific market. The grower is then paid on the weight of the live birds, the payment being dependent on efficiency factors such as feed conversion and on how his efficiency compares with that of neighbors who had flocks processed during the same week.

The first and most serious flaw in the relationship is the unilateral contract arrangement, one developed with no grower input. Development of the contract by one party without concern or input from the other party opens the door for abuse and manipulation. This exacerbates the second flaw, which is the lack of accountability for quality of inputs furnished by the processor. The grower has to meet contract specifications in every area of responsibility, yet the processor, who controls factors most critical to growth performance, has no accountability for delivering quality inputs. This is seen in the following contractual factors (see table 9.2):

Table 9.2
Contract Factors Affecting Pay Determined by Tyson

Factor	*Accountability*	*Effects and Framework*
Cents per pound paid		Varies based on ranking system
# of chicks placed		Varies with seasonal demand
Chick breed placed	None	Breed performance varies, affects pay
Sex of chick placed		Male/female performance differs
Health of chick placed	None	Grower must do best with that provided
Placement time	None	Nonpaying outtime varies with demand
Feed content	None	Caloric/nutritional content affects pay
Feed volume	None	Pay based on lb feed/lb gain ratio
Vaccinations	None	Bird performance directly affected
Medications	None	Bird performance directly affected
Market time	None	Age affects lbs added and conversion
Farm costs	None	Costs of chicks, feed, medication used in system to rank growers
Equipment additions		May be required whether grower benefits or not from investment
Housing modifications		May be required whether grower benefits or not
Dead bird disposal		May require more costly method than regulations permit
Processing condemnations		Grower may pay for plant mistakes
Net weight of birds		Scales periodically tested by state but weighing not witnessed
Time of weighing		Dehydration and weight loss affect pay
Who does labor	None	
When labor done	Yes	Requirements in grower handbook

These contracts result in payments ranging from three to six cents per pound for the live weight delivered. The variance in pay substantially relates to the quality of the inputs furnished by Tyson. A grower can employ the very best management practices on his farm and end up with the lowest pay if the chicks were sickly, if there was a higher ratio of female birds to male birds, or if the quality of feed received was poor, while his neighbors received better chicks, feed, and other input items. The grower's annual income will be reduced if fewer birds are placed per house or the downtime between flocks is increased as a result of excess broiler meat in the marketplace. These factors are all in the control of the company, and there is no accountability for the grower's needs in the arrangement.

Contract provisions specified by Tyson may require the expenditure of additional capital or the burden of additional operating expenses, whether there is economic benefit to the grower or not. The grower usually is left with no option if the company requests installation of new equipment or changing operating procedures which may increase farm costs.

An example of how these contract factors work is seen in Tyson's Grannis, Arkansas, complex. The company developed a market in the Far East for chicken feet and in 1995 wanted to improve the quality of bird feet by installing air cannons to ventilate the facilities. The cannons would cost several thousand dollars per farm. Feet were formerly ground, cooked, and mixed in the chicken feed, but the market would pay over eight times its feed value. No pay increase was offered to help offset their cost, yet growers were told there would be no more bird placement until the air cannons were installed.

In another example, Tyson had freezers installed on many of its contract farms in which to store dead birds before sending them to a rendering plant. The grower must pay the electrical costs of sixty dollars per month and up for operating the large freezers. Less costly methods of disposal, such as digestors, are available, yet Tyson can take the dead birds, render them, and produce and sell pet food products without sharing profits with growers.

Examination of factors over which Tyson has sole decision making power leads to the simple conclusion that the grower is at the mercy of the company. This conclusion is supported by the continuing decline in net income for poultry growers, the growing abuse reflected in the many complaints being recorded by the Packers and Stockyards Administration, and the growing number of individual civil actions against processors in various courts across the nation.

The Packers and Stockyards Administration (P&SA), now a part of the Grain Inspection, Packers and Stockyards Administration (GIPSA), was established through the Packers and Stockyards Act of 1921 to oversee the beef, pork, and lamb industries. The act prohibited the use of unfair and deceptive trade practices as well as addressing monopolistic developments in these areas. In 1987, the act was amended to include P&SA oversight of poultry and contract poultry growing arrangements. The P&SA has asked for budget increases to address the increasing number of complaints being filed by poultry growers.

Several class action suits challenging the use of deceptive and unfair trade practices by processors that affect thousands of growers have been filed. For example, in *Braswell v. ConAgra, Inc.* (936 F.2d 1169; 11th Cir. 1991) growers under contract with ConAgra in the Enterprise Alabama Complex received over $17 million in judgments and interest for fraudulent practices used in weighing live birds. Other cases are in mediation. Growers in northern Florida have cited violations both of the Packers and Stockyards Act of 1921 and of the Agricultural Fair Practices Act of 1967. And in *Wiles v. Tyson Foods, Inc.* (5"94cv4-v" w.d.N.C.) growers allege cheating on head counts, weighing of birds, and processing plant condemnation deductions. Trouble is brewing in the poultry industry as a result of the absolute power held by processors.

SWINE INDUSTRY DEVELOPMENTS

Though they occurred at different times, the swine and poultry industries have similar historical phases. The swine industry has provided the market's demand for pork with independent producers some forty years longer than producers in poultry. However, we can distinguish clear parallels between historical phases in the two industries.

Phase 1

Prior to the last decade, essentially all pork was owned and produced independently by farmers who took their pigs to auction. This phase corresponds to that in poultry when John Tyson was going farm to farm to buy chickens.

Phase 2

About ten years ago the corporate swine farm entered the picture, where huge numbers of pigs were grown and slaughtered by a single company. This is most notable in North Carolina where corporate farms and processing have driven most of the small auctions and slaughter plants out of business. This phenomenon closely parallels the period in which John Tyson began hatching and growing his own birds, essentially eliminating the market for growers from whom he had been buying. These corporate swine farms are operating in a number of states now, and the small independent farmer is finding it increasingly difficult to find a market. The tremendous production by mega-farms also drives the price of pork down over time, contributing to the demise of many independent farms.

Phase 3

Vertically integrated contract production begins to enter the swine industry in areas such as North Carolina where markets for independent producers have dried up. With pork price declines, contract swine farming is being introduced in several areas such as Kansas, Oklahoma, and Arkansas. This phase corresponds to that of the poultry industry in the sixties when most chicken production moved to contract farms. The change is characterized by the same factors as in the poultry industry, the primary factor being the use of farmer capital instead of corporate capital to expand. An interesting example is the economic projection by Tyson's Holdenville, Oklahoma, operation for a swine finishing farm. The producer invests $224,000 in two fully equipped confinement production houses in return for a $55,000 annual cash flow (if everything goes right) and a net profit of about $13,000. This is similar to building two tunnel ventilated poultry houses costing $225,000 that gross $50,000 to $60,000 annually with similarly low net profits.

Phase 4

This is the current phase of poultry industry development in which the processors have complete control over all aspects of the production process from embryo to the market shelf. The swine industry has not yet entered this phase, as processors have not established absolute control over producers. As more independent producers are eliminated from the business and the majority of production is done under contracts, difficulties in production will multiply. Pork producers will be situated similarly to poultry growers both economically and socially.

How can these conclusions be drawn so clearly? Three considerations lead to these conclusions: (1) the swine industry is moving swiftly to vertically integrated production as the primary means of meeting and creating market demand; (2) many of the major players in the poultry industry are heavily involved in pork production, such as Tyson Foods, Continental Grains (Wayne Farms is a poultry division), Seaboard, and Cargill; and (3) contract arrangements are almost identical in the pork and poultry industries.

The first two factors are discussed in many forums and require no further detail here. Discussion of contract similarities provides substantiation for the conclusions. An examination in table 9.3 of the Tyson Feeder Pig Producer Contract and comparing it to the broiler contract provides valuable insight into the swine industry's future.

As illustrated in these contract comparisons, the swine producer is responsible for many product quality factors under the contract, while the company has very few. This is the same story we see in poultry contracts. There is further examination in table 9.4 of swine contract provisions.

Table 9.3
Swine Quality Comparison to Poultry Contract

Contract Item	*Accountability*	*Same /Different*
Furnished by Tyson		
Contract language		X (no producer input)
Pigs	None	X
Feed	None	X
Medication	None	X
Vaccines	None	X
Management	None	X
Technical advice	None	X
Grower pay basis		X (pay based on wt. Gain)
Furnished by Producer		
Confinement houses	Yes	X (meet Tyson specs)
Equipment	Yes	X (meet Tyson specs)
Utilities	Yes	X (meet Tyson specs)
Labor	Yes	X (follow Tyson guide)
Carcass disposal	Yes	X (follow all regulations)
Manure disposal	Yes	X (follow all regulations)
Equipment updates	Yes	X (periodic inspections)

Table 9.4
Swine Contract Provisions

Contract Items Affecting Pay	*Determined by Tyson*	*Determined by Producer*
Pay per pound or head	X	
# of pigs placed	X	
Breed placed	X	
Placement date	X	
Feed content	X	
Feed volume	X	
Vaccinations	X	
Medications	X	
Market time/weight	X	
Equipment additions	X	
Housing modifications	X	
Carcass disposal	X	
Unmarketable pigs	X	
Net weight gain of pigs	X	
Who does labor		X
When labor done		X (meet Tyson guideline)

The most serious flaw in the swine contract relationship is the unilateral contract, or one developed with no producer input. This is the same problem that is well documented in the poultry industry. The second serious flaw is the lack of accountability for quality of inputs furnished by the processor. Swine producers can accurately use the poultry industry as a means of gauging what their future holds. The commonality of involved companies, industry structure, contract form, and production techniques can lead in no other direction as long as producers accept what is offered.

CONCLUSION

Swine producers do not have to end up in the same struggle as poultry producers. When vertical integration began to develop in the poultry industry, there were no poultry producer organizations to look after the growers' interests. In fact, companies strongly discouraged grower organizations. Passage of the Federal Agricultural Fair Practices Act of 1967, prohibiting the use of unfair and deceptive trade practices because of an individual's participation in a growers' organization, was passed to a large extent because poultry companies were terminating and blackballing growers who attempted to organize. Swine producers already have several organizations in place across the country. The challenge of contract farming may be met with the development of a national voice for pork producers such as the poultry growers are working to achieve through their agricultural cooperative, the National Contract Poultry Growers Association.

REFERENCES

Atlas, Riva. 1994. Food, drink and tobacco. *Forbes* 153(1): 152–54.

Clouse, Mary. Farmers net income from broiler contracts. Rural Advancement Foundation International—USA, January 1995.

Gallagher, Peter. Economic returns for U.S. poultry growers. National Contract Growers Institute, May 1995.

Heuslein, William. 1995. Food, drink and tobacco. *Forbes* 155(1): 174–76.

Part IV

Alternatives

A basic point of this volume is to better understand our industrial system of food production in order to better articulate alternatives for the future. The two contributions in this section point to alternatives for the future. In contrast to considerable rhetoric prophesying the inevitability of forces driving industry changes, these contributions clearly show that we do have options. Humans are malleable cultural creatures who can learn from the past to create our future. However, cultural evolution, in contrast to biological evolution, is based on experiments of change brought about by ideas and visions concerning the way we adapt to our environment. A movement known as "sustainable agriculture" is harkening in a number of experiments with new food production systems, including a growing number of organic producers and community supported agricultural systems whereby food producers and consumers are directly linked. The two contributions in this section integrate a pragmatic understanding of farming and farm economics with suggestions for new ways of thinking about food production.

John Ikerd, an agricultural economist, blends an understanding of farm economics and community development with a philosophical vision for a new sustainable agriculture. Ikerd provides a detailed critique of a typically narrow economic impact assessment of large-scale swine operations in Missouri. He points out that industrializing hog production is by its very nature intended to replace workers with technology and consequently will have a negative net affect on employment. Ikerd broadens the scope of thinking by moving beyond the economic numbers by identifying and addressing the assumptions and philosophies that underlie them in order to formulate a new way of thinking about food production.

In his contribution, Randy Ziegenhorn, a farmer and anthropologist, describes how cooperative networks of independent pork producers can form as an adaptive response to pressures from large-scale corporate producers. Based on his ethnographic work among organizers and members of pork producer networks in Iowa, Ziegenhorn cracks the code for successful producer networks. Along the way, he points out the shortcomings of business-oriented approaches to producer networks promoted by agricultural colleges that fail to take into account the social reality of independent pork producers.

Chapter 10

Sustainable Agriculture, Rural Economic Development, and Large-Scale Swine Production

John E. Ikerd

INTRODUCTION

In late 1994 increasing pork supplies dropped hog prices to levels that were unprofitable even for the most efficient hog producers. After a modest recovery during the winter, hog prices again fell to unprofitable levels in spring 1995. Through all these ups and downs, large corporate hog operations continued to expand, displacing independent producers with each downturn in prices. There is little doubt that a large number of independent hog producers will continue to be displaced as corporate operations continue their expansion.

The expansion of large corporate hog operations made these hog cycles fundamentally different from cycles of the past. Hog farmers who fail to recognize this difference, who just try to ride out cycles as they have in the past, may well face serious financial stress and a high probability of failure. Independent producers, such as Jim Braun, who also has a chapter in this book, who choose to remain independent must develop a system which capitalizes on their ingenuity and creativity if they are to survive the demands and interests of large, vertically integrated hog operations. Sustainable agriculture may offer smaller, independent hog farmers hope for finding an alternative to large-scale industrial production.

SHIFTING PARADIGMS AND SUSTAINABLE AGRICULTURE

A paradigm establishes or defines boundaries and sets standards for success and behavior within the boundaries (Barker 1993). Barker uses the game of tennis as an analogy to illustrate these concepts. Tennis courts are a standard size and their boundaries are clearly marked. The ball must hit within these boundaries to stay in play. The ball must be struck with a tennis racket, not a baseball bat or anything else, and the ball is allowed to bounce only once before it is returned over the net.

Paradigms may be simple, as in the case of games, or extremely complex, as in the case of a model for economic development. However, the industrial model has clearly defined boundaries. The natural environment and resource base, as well as community social issues, are considered external, or out of bounds, by industrial managers. Success for an industrial firm is measured in terms of profits and growth. Industrial firms engage in a wide range of actions to maximize short-run profits and long-term growth. Virtually anything that is possible and legal is encouraged if it leads to profits and growth.

Our current industrial model or paradigm for farming may not be sustainable over time. No one knows the future with certainty. It may be possible to fine-tune, refine, or redesign the industrial model, resulting in a new model that will meet the ecological and social standards required to sustain long-run productivity. However, an approach or philosophy of farming that is fundamentally different from the industrial model may be required. This alternative paradigm may represent the independent hog farmer's best hope for surviving in the face of large-scale corporate hog production.

The new sustainable agriculture paradigm may not be fully understood for some time. However, this postindustrial approach to farming is fundamentally different from the industrial paradigm in several ways. The new model for agriculture clearly considers ecological and social impacts to be within rather than outside its boundaries. Ecological soundness and social responsibility are seen as positive goals rather than negative constraints. The new paradigm considers economic, ecological, and social dimensions of sustainability to be inseparable. Fields, farms, and communities are wholes made up of smaller systems which constitute still larger wholes. Thus, the approach to farm decision-making and management must be "holistic." The challenge is to comprehend the complexities of wholes rather than reduce them to simple elements. Success in the new model is measured against the goal of sustainable economic, ecological, and social progress rather than profits and growth.

A new sustainable model relies more on people, including the quality and quantity of labor and management, and relies less on land and capital. Success of the new model for farm decision-making may well depend on empowering more people with information and knowledge needed for holistic management. Table 10.1 outlines points of comparison between industrial and sustainable paradigms for farming. These differences cannot be defined in two specific sets of production practices or methods, but instead must be considered as contrasting approaches, philosophies, and policies that guide farming.

A fundamental shift in the balance of returns to people versus land and capital must be brought about if the new paradigm is to be successful. Smaller, diversified farms will survive alongside larger, specialized farms only if human resources can be substituted economically for other resources and commercial inputs. Farmers who succeed with the new model must be rewarded as more productive people contributing to our social, political, and economic environments.

Rural community development strategies are also undergoing a significant change consistent with a new paradigm of economic development. The old strategies of industrial recruitment through building industrial parks by offering tax breaks has given way to growth-from-within policies. The new strategy, in line with the business theories of Reich (1991), Toffler (1990), Drucker (1989), and Naisbitt and Aburdene (1990) invests in entrepreneurs within the community to build small businesses and strengthen the local economy. Local

Table 10.1
Alternative Paradigms for Farming

Industrial	*Sustainable*
Reductionist (seeks to simplify)	Holistic (seeks to comprehend)
Complex Systems (interconnected parts)	Complex Wholes (wholes within wholes)
Inductive Reasoning (parts⇒whole)	Deductive Reasoning (wholes⇒"parts")
Separable Goals (multiple goals)	Inseparable Goal (multidimensional)
Focus on Productivity and Profits	Focus on Sustainable Progress
Ecological and Social Constraints	Ecological and Social Objectives
Economic, Ecologic, Social—Separable	Ecologic, Economic, Social—Inseparable
Constrained Profit Maximization	Economic, Ecological, Social Balance
Strategy: Linear/Sequential	Strategy: Spiral/Simultaneous
Values Conformity/Specialization	Values Uniqueness/Diversification
Driven by Problem Solving	Driven by Problem Prevention
Technology Based (applies knowledge)	Knowledge Based (applies technology)
Stresses Knowledge of Facts	Stresses Human Ingenuity
Management Extensive (manage more)	Management Intensive (more managers)
Management by Objectives	Management by Principles
Promotes Sound Management Practices	Promotes Disciplined Use of Judgement
Tests Decisions against Objectives	Tests Decisions against Goal
Monitors Objective-Based Results	Monitors Impacts on Whole System
Balances Independence/Dependence	Balances Independence/Interdependence
Quality of Life Is a Result of Success	Quality of Life Is the Process of Success
Motivates People	Empowers People

buyer-supplier projects replace dollars leaving the community from outside corporations with locally produced goods and services.

Efforts to attract low-quality, low-paying jobs are increasingly regarded as an expensive and ineffective strategy for economic development as Mark Grey shows in his chapter. Large companies may provide a large number of jobs, but they often pay poorly and may be unstable since there is no local commitment, they are expensive to attract and maintain, they are slow to respond to new economic conditions, and they bring social and environmental costs which tax payers must bear. Economic development professionals are beginning to concentrate on improving the quality of jobs rather than increasing the quantity.

The sustainable agriculture paradigm supports a growth-from-within approach to rural economic development. It is an asset-based rather than deficiency-based strategy where human resources are more highly valued than financial ones. As intellectual assets are employed, they are enhanced rather than depleted as with other resources. Sustainable development offers an alternative to the vicious cycle of industrial recruitment, low wages, declining emphasis on education, declining communities, and resulting downward spiral (Reich 1991). This alternative is the renewable cycle of education, increased innovation, increased investment, increased value, and higher wages. Sustainable agriculture relies on knowledge of the land and ecology that supports the production of food and fiber. The strength of this approach lies in human intellectual capacity to work with nature and maintain productivity. It focuses on empowering farmers and local communities through dependence upon people rather than money.

IMPLICATIONS FOR PORK PRODUCTION

As John Morrison discusses in his chapter, the past may provide a picture of the future of midwestern family farms and rural communities as hogs now take the lead as the latest form of agricultural industrialization. Hogs constitute an important sector of the midwestern agricultural economy. What is more important, hogs are the economic backbone of many diversified family farms.

Hogs provide a value-added market for crops, stabilized farm income, recycled nutrients, and scavenged wastes. Hogs and other livestock provide year-round employment for farm family members who are otherwise underemployed in specialized cropping systems. Hogs on farms provide a classic example of synergistic productivity. They are capable of adding far more to farming systems than is apparent from superficial economic examinations of hog enterprises such as the one described below.

A 1992 University of Missouri report begins with the following statement: "The State of Missouri is poised at a crossroads with respect to pork production. Developments within the state in the next five years will likely determine the future of pork production in Missouri well into the twenty-first century" (DiPietre 1992:73). This report leads one to believe that the major problem confronting Missouri's hog producers and rural communities is a drop in total pork production. The solution offered to this problem is to bring in large, multistate corporate hog producers to increase the number of pigs Missouri produces. Much of the report focuses specifically on the potential positive economic impacts of increasing large-scale, contract swine production in Missouri. A central but unanswered question in this report is whether increasing contract swine production, or any system of large-scale hog production, is a sustainable rural economic development strategy for Missouri or any other state.

The fundamental challenge to the future of Missouri's farms and rural communities is the declining availability of quality employment opportunities. Only to the extent that it results in higher quality employment opportunities does increased hog production contribute to economic development of rural communities. Since large-scale confinement hog operations seek to reduce total costs by using production methods that allow fewer people to produce more hogs, the substitution of capital and mass-production technologies for labor and management is the primary advantage that large, specialized hog production units have over smaller, diversified operations. The displacement of family farms by large-scale hog production systems reduces rather than increases total employment in pork production.

The production environment in large-scale operations is controlled through utilization of buildings and equipment that require large capital investments but greatly reduce labor requirements. Production technologies associated with large-scale, contract production also change the nature of management. Mass-production technologies that standardize genetic selection, breeding, feeding, herd health, and marketing functions transfer most of the management function from on-site hog producers to corporate production supervisors who travel among production units and to production managers at corporate headquarters. Large-scale, specialized hog production replaces people with capital intensive, mass-production technologies and centralized management.

A comparison of results of the Missouri contract production projection (DiPietre 1992) with reports from actual Missouri hog operations illustrates the basic principle of input and resource substitution. Figures in the first three columns in table 10.2 show contract production data taken from the DiPietre study. The first column shows data for a basic six hundred sow contract farrowing unit producing twelve thousand pigs per year. A $550,000 investment is expected to result in 2.5 on-site jobs with a total direct employee compensation of $22,800 per job. The assumed 2.22 impact multiplier results in a total of 5.55 on-site and off-site jobs.

The second column in table 10.2 shows comparable data for 3.5 contract hog finishing units that feed out hogs to market, a number sufficient to produce

Table 10.2
Comparisons: Contract and Independent Hog Production

	Contract Farrowing	*Contract Finishing*	*Contract Farrow-to-Finish*	*Hog Management Information Records (MIR)*
Production Units	1.00	3.50	1.00	4.50
Sows	600	0	600	
Pigs/year	12,000	11,900	11,900	
$Investment B&E	550,000	455,000	1,005,000	1,192,500
$Total Assets	550,000	455,000	1,005,000	2,452,500
$Sales		1,368,500	1,368,500	1,306,071
Cwt-Sales			28,510	27,210
Mgt-Labor	0.00	0.00	0.00	6.17
Other Labor	2.50	1.75	4.25	6.44
Total Labor	2.50	1.75	4.25	12.60
Jobs Displaced			8.35	
$Assets/Person			236,471	194,643
$Return: Mgt+Capital	**	**	**	270,332
$Return: Labor	57,000	36,750	93,750	78,975
$Return: Mgt+Labor	57,000	36,750	93,750	349,307
$/Unit-Mgt+Lab+(Cap)	22,800	21,000	22,059	27,723
Multiplier	2.22	2.22	2.22	2.22
Total Employment	5.55	3.89	9.44	27.97
Jobs Displaced			18.54	
Displacement Ratio			2.96	to one
$5 Mil. (9 & 11 units)	50.00	42.79	47.18	139.86
Jobs Displaced			92.69	

11,900 hogs per year, comparable to the number of pigs produced by the one basic contract farrowing unit in the first column. In this case the investment results in a total of 1.75 on-site jobs at an average salary of $21,000. The total employment associated with 3.5 finishing units is 3.89 jobs.

The third column represents a composite of columns one and two. The total investment is approximately $1 million, including buildings and equipment costs but no land. Total dollar sales are based on 11,900 hogs sold at 250 pounds for $46 per hundred pounds. The number of hogs and price is the same as those assumed in the original study. A market weight of 250 pounds is the average reported in 1992 Management Information Records (MIR) production data for hog finishers (Plain 1993). The composite operation results in an estimated 4.25 on-site jobs at an average salary of $22,059. There is a total of 9.44 jobs that includes off-site employment.

The fourth column is based on actual production data reported in the 1992 *Missouri Farm Business Summary* (Ehlmann and Hein 1993). Twenty-five farms participating in the MIR program in 1992 were classified as hog farms. Sales of livestock accounted for all positive net returns on these farms. In fact, crop returns showed a net average $12,000 loss to operators. Slaughter hog prices in 1992 averaged about $42 per hundred pounds, $4 per hundred pounds less than the average used in the contract units in the first three columns of table 10.2. Adjusting 1992 sales upward to reflect the $4 price difference, total sales for 4.5 MIR hog farms are estimated at $1.3 million. This brings total sales for MIR hog farms in column four roughly in line with sales for the composite contract farrow-to-finish unit in column three.

A key difference between contract and individually owned hog production is in management and labor requirements. The composite contract farrow-finish contract operation in the third column employs only 4.25 people to generate $1.3 million in hog sales. In contrast, independently operated hog farms employ 12.60 people. Large-scale, specialized operations create fewer jobs per hog produced but more hogs per employee.

It is estimated that 2.22 jobs are created indirectly for each person directly employed in hog production. Thus, the total number of people needed to produce twelve thousand hogs is estimated at 9.44 persons for corporate contract production and 27.79 persons for independent hog farms. Corporate contract production replaces approximately three independent hog farmers for each new job they create. Large-scale contract production employs far fewer people than would be employed to produce the same number of hogs in typical owner-operated hog farms. If new contract hog units were to replace independent operations producing the same number of hogs, the Missouri study indicates that approximately two hog farmers would be left without jobs for each new job created. In the process of changing $1.3 million in hog production from independently owned production to contract production, a net of 18 to 19 jobs currently linked to hog production would be displaced. A new $5 million investment in contract production would generate 40 to 50 new jobs but would displace approximately three times that number of independent hog farmers.

Some of the difference in employment occurs because many hog farmers produce a significant portion of their own feed, whereas contract operators typically purchase their feed from outside suppliers. Management functions performed on the farm by independent hog farmers are performed by off-farm supervisors and corporate managers in contract operations. However, when feed is produced on the farms where it is fed and returns to the management of local farmers, there is little doubt about the positive economic impact in the local community. A recent Minnesota study (Chism and Levins 1994) examined tax records of large and small livestock producers and concluded that smaller producers bought a greater proportion of their inputs locally. For smaller livestock farmers (annual gross sales less than $400,000), the weighted average of local spending (within twenty miles) was 79 percent, whereas for larger farmers the weighted average of local spending was only 47.5 percent. These data indicate that smaller to moderate-sized hog operations are more supportive of local input suppliers than either larger owner-operated or corporate operations. Moreover, corporate or contract producers might have fewer ties to the local community and even less loyalty to local suppliers than larger independent producers.

As also indicated in table 10.2, total assets per person employed in contract production is higher than for independent hog farms, even though land needed for the contract unit is not included. Total investments on independent farms include all land, machinery, and equipment associated with all enterprises, including on-farm feed production. However, the contract production asset figures are based on new investment costs, whereas facilities on independent hog farms are of varying ages and are reported at depreciated values. Thus, asset comparisons in table 10.2 can be misleading because they reflect a reality in 1992 before contract facilities depreciate.

The DiPietre report (1992) includes only direct employee compensation, or returns to labor, in estimates of economic performance for contract units. Direct employee compensation for contract production also includes the combination of interest and principal payments on buildings and equipment. If the useful life of buildings and equipment does not exceed the repayment period, additional returns to the contract producer's investment will decline as indicated in John Morrison's chapter on poultry facilities in this book. Total returns to management, labor, and capital for contract operations in table 10.2 reflect only direct employee compensation estimates.

Returns to capital, management, and operator labor are fundamentally inseparable in owner-operated farming enterprises, although separate returns for different production factors are often estimated. Estimated returns to management, capital, and labor per employee is over 25 percent higher for independent hog farms in table 10.2 than the direct employee compensation per person in contract operations. Some of this difference is likely due to the fact that returns to capital accrue to the farmers in independent hog operations, but accrue to corporate managers and investors in contract operations.

Specific impacts of farm size and ownership structure on local retail and service sectors are not directly captured by conventional economic impact assessment models such as the one used in the DiPietre report (Ikerd, Devino, and Traiyongwanich 1996). Data for conventional models, such as Implan, are based on relationships between incomes of hired workers and total consumption transactions averaged over the total local economy. Much of the difference in total economic impact between large-scale corporate and smaller family-based systems of farming stems from differences in personal, spendable income. Returns to owned land, equity capital, and family labor accrue as spendable family income in family farming operations. Returns to these factors appear as payments for rents, interest, dividends, and hired labor in most corporate businesses.

For family farms, total returns over direct production costs are available for local retail spending. For most corporate operations, only those costs paid in local wages and salaries are available for retail spending. Using average income figures tends to underestimate the impacts of owner-operated family farms on retail employment and overestimates their impacts on employment in manufacturing and wholesaling. However, relatively few agricultural inputs are manufactured in rural areas. In addition, fewer people are typically employed per dollar of sales in wholesaling compared to retailing. Thus, average impact multipliers tend to underestimate the potential impacts of family farms on local incomes and employment. Successful owner-operated family farms tend to be smaller and, in general, employ more people per acre or per dollar of output than do corporate enterprises. More people in the local community tend to generate more local retail spending.

A Nebraska study compared detailed economic data provided by twenty-eight farmers, half of which were classified as "conventional" and the other half as "sustainable," based on their current farming methods (Kleinschmit, Ralston, and Thompson 1994). Farms identified as "sustainable" were about one-half as large in terms of acres, head of livestock, and total sales, as those labeled "conventional." However, sustainable farmers actually reported a higher average farm income, or return over direct costs per farm, in spite of their smaller size. A total of 169 people were supported by the 28 farms included in the Nebraska survey. It was estimated that an additional 44 people could have been employed on the same number of acres with at least as high a per capita income if all farms in the survey area had been of the same average size as the sustainable farms. If all farms were like those in the conventional group, 22 fewer people could have been supported in the area, a difference of 66 people. Total family income with all sustainable farms, including on-farm and off-farm income, would have been more than double that of an all conventional community and 80 percent higher than in the current community. A typical economic input/output impact analysis, however, would conclude that the two types of farms have very similar impacts on the local community because they generate similar levels of gross agricultural sales.

The obvious local argument in favor of locating contract hog production in any community or state is that the gains will be concentrated locally and the losses will be spread elsewhere. A more common argument is that if farmers in one state or region do not expand through corporate or contract production, producers in other states will, displacing those who do not participate in the trend. Comparative data from North Carolina and Nebraska provide evidence to the contrary. Nebraska passed a state constitutional amendment restricting corporate farming, while North Carolina is a national leader in promoting corporate hog production. During the 1986 through 1993 period, while the number of hog farmers in North Carolina fell by nearly one-half, total hog production and the total numbers of hog farmers in Nebraska remained virtually unchanged (Center for Rural Affairs 1994).

Nebraska has kept its hog farmers employed without promoting large-scale corporate contract hog production. However, the farmers displaced most frequently by increased corporate hog production in North Carolina have been independent North Carolina hog producers. As Thu and Durrenberger point out (1994), negative economic and social impacts have been felt most keenly in North Carolina's rural communities. This evidence suggests that while the gains from contract production may be concentrated close to home, economic losses may be felt nearby as well. The pain of displacement and dislocation cannot be exported to other states.

Comparing independent hog farms to contract hog production units is like comparing apples to oranges. One represents whole-farm operations, the other represents specialized hog enterprises. In one, the equity and management is on-farm, in the other most of the capital and management comes from the outside. One set of assets represents an investment in a specific set of buildings and equipment, the other represents a wide variety of capital assets. So how can such comparisons have any real meaning? The comparisons have meaning precisely because the two systems represent two very different but real futures for hog production. One alternative is to promote large-scale, contract production as a means of increasing hog production in hopes of retaining or regaining employment. Another strategy for the future is to develop a new paradigm for pork production. These strategies require very different kinds of policies, as well as financial and human resources.

Hog producers have a variety of potential paradigms for hog production in the twenty first century. To support only one model of twenty first-century hog production is to imply that the only choice is between apples and nothing. Hog farmers and others who live and work in rural communities should be allowed to choose their own future. If they choose apples over oranges, it is their choice. However, given a choice of apples, oranges, and bananas, they just might choose some of all three. To be meaningful, such choices have to be freely made and based on reliable information from a variety of perspectives.

IMPLICATIONS FOR THE FUTURE

Corporate-owned and contract hog production are realities of the current agricultural economy. It seems quite likely that it will become more common in spite of attempts by opponents to slow its growth. Industrialization at first appears to increase productivity of human resources through mechanization and mass production technologies. However, over time, industrial technologies replace more and more labor with mechanization and the management function becomes concentrated among fewer and fewer people.

The traditional economic assumption has been that more efficient production will increase total economic output to provide new jobs for displaced workers in newly emerging sectors of the economy. However, the growing number of displaced workers in the American economy, who now range from underemployed to permanently unemployable, raises serious questions about these traditional economic assumptions. The productive employment of people may be the most fundamental economic and social problem that the industrial model is inherently incapable of solving.

Rural America cannot retreat to an earlier time when cost competition was less keen and full-time family farms were the norm. However, there is clear evidence that independently owned, modest-sized, family-operated hog farms can be competitive with large-scale contract production units. A new paradigm for farming from a sustainable agricultural framework supports the goal of economically viable, ecologically sound, and socially responsible systems of farming. Social responsibility includes a commitment to providing quality employment opportunities on independently owned, small to moderate-sized, family farms. Moreover, as Durrenberger and Thu (1996), and others (Barnes and Blevins 1992; Goldschmidt 1978; Lobao 1990; Office of Technology Assessment 1986) point out, there is clear evidence that successful, modest-sized, family-operated hog farms contribute more to the economic and social well-being of rural communities than do their corporate counterparts.

A goal of improving the management capabilities of independent hog farmers should receive at least as high a priority as contract hog production as a rural economic development strategy. The goal of developing a new sustainable paradigm for agriculture, including hog production, may be even more important in the long run. Greater reliance on intensive management creates more quality employment opportunities in rural areas by enhancing the productivity of people, rather than replacing people with capital investments and mass-production technologies. Empowering people to be productive is the foundation upon which the new paradigm of sustainable agriculture must be built.

Some will demand concrete evidence of the success of this new paradigm before they will act. According to Barker, typical responses to those who propose new paradigms include: "That's impossible." "It's too radical." "We tried something like it before, and it didn't work." "I wish it were that easy." "Let's get real, okay?" How dare you suggest that what we are doing is wrong!" One may look to the decentralizing and deindustrializing of the rest of the U.S.

economy for evidence that the industrial era is coming to an end. However, those "who choose to change to a new paradigm early must do so as an act of faith rather than the result of factual proof, because there will never be enough proof to be convincing in the early stages" (1993: 199).

Those who wait until the value of a new paradigm can be proven will lose the advantage to those with the courage to change in the face of uncertainty. Independent hog producers do not have the luxury of waiting to see if something else will work. They must find a new paradigm for farming soon, or they will soon be doing something other than farming. The fundamental question is whether they can expect real help from the academic and scientific community in meeting this challenge, or whether they will be advised to accept the "inevitability" of industrialization.

REFERENCES

Barker, Joel. 1993. *Paradigms: The business of discovering the future.* New York: Harper Collins.

Barnes, Donna, and Audie Blevins. 1992. Farm structure and the economic well-being of nonmetropolitan counties. *Rural Sociology* 57: 333–346.

Center for Rural Affairs. 1994. *Corporate farming update! Spotlight on pork.* Special Report. Walthill, Nebr.: Center for Rural Affairs.

Chism, John W., and Richard A. Levins. 1994. Farm spending and local selling: How do they match up? *Minnesota Agricultural Economist* 676: 1–4.

DiPietre, Dennis D. 1992. The Economic Impact of Increased Contract Swine Production in Missouri, Missouri Farm Financial Outlook 1993, University Extension, University of Missouri-Columbia, Department of Agricultural Economics.

Drucker, Peter. 1989. *The new realities.* New York: Harper and Row Publishers, Inc.

Durrenberger, E. Paul, and Kendall M. Thu. 1996. The expansion of large scale hog farming in Iowa: The applicability of Goldschmidt's findings fifty years later. *Human Organization* 55(4): 300–308.

Ehlmann, Gerald M., and Norlin A. Hein. 1993. Missouri Farm Business Summary, 1992. University of Missouri-Columbia, Extension Division, FM 9390.

Goldschmidt, Walter. 1978. *As you sow: Three studies in the social consequences of agribusiness.* Montclair, N. J.: Allanheld, Osman.

Ikerd, John E., Gary Devino, and Suthijit Traiyongwanich. 1996. Evaluating the sustainability of alternative farming systems: a case study.

Kleinschmit, Linda, Don Ralston, and Nancy Thompson. 1994. Community impacts of sustainable agriculture in northern Cedar County, Nebraska. Special Report of the Center for Rural Affairs. Walthill, Nebr.

Lobao, Linda. 1990. *Locality and inequality: Farm and industry structure and socioeconomic conditions.* Albany: State University of New York Press.

Naisbitt, John, and Patricia Aburdene. 1990. *Megatrends 2000.* New York: Avon Books.

Office of Technology Assessment. 1986. *Technology, public policy, and the changing structure of American agriculture.* Washington, D.C.: Office of Technology Assessment.

Plain, Ron. 1993. 1992 Missouri Hog Returns. Farm Management Newsletter, Department of Agricultural Economics, University of Missouri-Columbia.

Reich, Robert B. 1991. The real economy. *The Atlantic* 267: 35–7.

Toffler, Alvin. 1990. *Power shifts.* New York: Bantam Books.

Thu, Kendall, and E. Paul Durrenberger. 1994. North Carolina's hog industry: The rest of the story. *Culture and Agriculture* 49: 20–23.

Chapter 11

An Alternative Model: Swine Producer Networks in Iowa

Randy Ziegenhorn

INTRODUCTION

On a cold February night in 1995, I sat in a nearly empty junior college classroom listening to a satellite teleconference on networking sponsored by the National Pork Producer's Council (NPPC). The teleconference, nationally broadcast from Des Moines, Iowa, featured a panel of experts and video clips of pork producers discussing cooperative business ventures. The moderator assured us that hundreds of farmers across the Midwest were participating in the teleconference. All of them were sitting in rooms like ours watching television screens where speakers urged them to form cooperative networks in order to survive sweeping swine industry changes.

The overriding purpose of the cooperative network is to improve the operations of its members in a way they could not as individual farmers. Among the possible forms of networks are information sharing, joint purchasing, joint marketing, and joint production. All are based on function, what the network does, rather than how the network does it. Although networks come in all shapes and sizes, from informal coffee shop discussion groups to corporations with multi million-dollar facilities, the message from the NPPC that night clearly emphasized the larger formal organizations. In place of the "mortgage lifter" sideline that hogs provided to traditional family farmers, hog farming is promoted as a specialized business enterprise operated by professional managers and driven by economies of scale and the global demand for pork.

"Change is coming," prophesied Glenn Keppy, a farmer and past president of the NPPC. The message from the expert panel of economists, livestock specialists, and producers emphasized such change and promised a bright future for those willing to embrace it. The experts and producers alike spoke the language of business schools— "networking is the end product of a strategic plan," intoned one, "Total Quality Management" was invoked by another, and a third encouraged producers to develop "mission statements." All of the solutions were presented as parts of programs to enable farmers to compete successfully in an industry increasingly dominated by large-scale corporate interests.

If there was a hitch to the optimism, it was summed up by a farmer from Illinois: "Pigs are easy; people are a challenge." Several speakers commented that "people problems" would interfere with the best laid plans of managers and experts. An economist described the failure of similar networks two decades earlier due to the inability of members to get along.

The NPPC teleconference represented the best efforts of academics, farm leaders, and producers to encourage a new kind of farming. As an anthropologist I wanted to find out whether this version of reality bore any resemblance to the actual experiences and ideas of pork producers involved in cooperative production networks. From 1995 to 1996 I studied the problems and prospects of networking among hog farmers in rural Iowa. What I saw convinced me that networks have the potential to benefit farmers, but not in the manner envisioned by planners who design networking strategies from office suites and lecture halls. If there is a people problem, it is the people who work from the assumptions of the business schools rather than the realities of farmers.

FARMER COOPERATIVES

The notion that farmers can better themselves by cooperating with one another may be as old as farming itself. Anthropologists have described the communal nature of food production and distribution in a wide variety of cultures. The reciprocal sharing of labor in food production occurred among our hunter-gatherer ancestors and still occurs in many peasant societies today. But the idea that farmers and other small producers in industrialized nation states could gain economic clout by banding together has its origins in the cooperative movement of the nineteenth century.

In the nineteenth century, farmers in the United States and Europe formed various kinds of "cooperative societies" to oppose centralized control from merchants, railroads, grain buyers, and meat packers. Operated as a form of collective bargaining, cooperatives attempted to curtail the market power of large firms by uniting the buying and selling of individuals under the collective clout of the cooperative. While one farmer may not affect the price of wheat, ten thousand farmers agreeing to withhold their crops from the market may.

Membership in cooperatives traditionally has been limited to farmers themselves, to those who buy from or sell to the cooperative. Cooperative decision making generally follows the principle of one member, one vote rather than the one share, one vote rule of corporations. In other words, no matter how much money he spent with the cooperative or how large his farm, the voting power of the individual member was no greater than that of any other member. This form of economic democracy appealed to many and contributed to the success of cooperatives and their popular reputation for protecting the interests of the little guy (Rasmussen 1991).

Cooperatives attracted a strong following in many parts of the United States, particularly among northern European immigrants to the upper Midwest. Today cooperatives handle much of the interior movement of feed grains in the upper Midwest as well as significant quantities of farm supplies, such as fertilizer and pesticides. Many local cooperatives joined to form regional associations which count their sales in the hundreds of millions. These large cooperatives became sophisticated businesses able to access capital and conduct trade on a national and international scale. Cooperative managers argue that they provide a force that brings economic benefits to their members that would not exist if privately owned firms dominated agriculture. Critics of cooperatives argue that as cooperatives grew in size and complexity they lost their populist appeal and are now just another form of corporate agriculture (Hightower 1972; Kravitz 1974).

In the 1970s, farmers in Iowa began experimenting with cooperatives to produce feeder pigs—weaned pigs weighing about 40 pounds ready for a feeding program to take them to market weight. The idea was to build a large centralized facility—a farrowing building—where the pigs would be born, hire a manager and employees to operate it, and distribute pigs to member farms for finishing. By the end of the decade there were over 80 such cooperatives in Iowa. These cooperatives were organized not as profit making businesses, but to distribute feeder pigs to members at the cost of production (Paulsen and Rahm 1979). By the 1990s few of these cooperatives remained. Many participant farmers described disagreements over management decisions and poor performance as reasons for the failure.

Large cooperatives traditionally avoided the actual production of agricultural commodities and livestock. Recent efforts by Farmland Industries, a large regional cooperative with its own packing operations, to fatten its own hogs has drawn sharp rebukes from the farm community (Des Moines Register 1994). Farmers expressed resentment that an organization which they ostensibly own is also a competitor. Farmland, like other large cooperatives such as Land O' Lakes and Growmark, responded to the growth of large corporate swine operations which emerged in the late 1980s. Unlike the small family farmer who buys feed at his local cooperative, these integrated firms mill their own feed, bypassing the cooperative or the privately owned feed company.

In an effort to adapt to these changes, many in the feed industry and land grant universities have proposed a new twist on the farmer cooperative—the network. In the social sciences, the use of the term *network* was coined by an anthropologist in the 1950s to describe an informal arrangement between various individuals for economic activity (Barnes 1954). A network represents various ties of kinship, friendship, employment, and patronage among individuals. By the 1980s network analysis became a burgeoning part of academic anthropology and sociology. The term *network* was transformed into a verb (as in "to network") by popular business writers who advocated networking as a means of strengthening businesses. In a sense, all cooperatives are networks, just as all businesses, schools, churches, and families are networks of ties and affiliations.

The use of the term *network* by pork producers today carries many of these overtones but generally refers to the cooperative-like efforts of several farmers to establish some form of joint venture. Perhaps the most important question facing many hog farmers today is whether they will direct the development of these new networks or whether these networks will assume a corporate form that directs them.

WHY NETWORKS NOW?

A number of factors emerged in the 1990s to rekindle interest in cooperation among hog farmers. Many small towns and rural communities fear the effect of a continuing loss of hog producers and the dollars lost to local economies by such an exodus. Regional feed companies and cooperatives with ties to family farmers fear the effect of dwindling numbers of hog producers. Others have a less direct stake. The leading national advocate for networks is the National Pork Producers Council, with strong support from economists and livestock scientists at land grant universities. The very existence of these groups is questionable if their constituency disappears. In addition, veterinarians, bankers, and politicians all support networking. Many network organizers even report encouragement from meat packers who fear that dwindling numbers of producers will mean greater difficulty in obtaining hogs for slaughter.

The consensus from NPPC and land grant universities is that small producers must employ the approaches of the largest producers if they are to compete. The problem seems to lie in convincing a specific farmer to join forces with others. One would-be organizer pointed to a hefty tome on networking from Iowa's Cooperative Extension Service and lamented, "This tells me everything to do to get a network going. It just doesn't tell me how to convince the farmers to join." In listening to organizers, extension workers, and farmers across Iowa a constant refrain is heard—farmers are too independent, too much enamored of the go-it-alone philosophy of the yeoman, to ever agree on anything much less cooperate with one another. That is the nature of the people problem, the essential "cussedness" of the independent family farmer.

I observed and interviewed farmers who decided to join a network rather than remain independent. Their decisions were based on a variety of factors—not the rational decision making that the economist would expect, but a very considered process nonetheless. Understanding how decisions are made to join a network is crucial if networks are to succeed as a sustainable alternative. Below I describe the experiences of two networks that shed light on how they might best be organized and sustained.

THE ETHNOGRAPHIC REALITY OF NETWORKS

During 1995 and 1996 I attended meetings of networks, conducted interviews with farmers and network organizers, and worked on a variety of hog farms in eastern Iowa. This process of collecting data known as "ethnographic fieldwork" is the hallmark of sociocultural anthropology. It represents the anthropologist's commitment to observing and participating in the daily lives of people in order to understand the world from their point of view. Unlike phone or mail-out surveys and other detached approaches, an anthropologist doing ethnographic fieldwork enmeshes himself or herself in the social and economic life of the people to see the world from a "grassroots" or "bottom-up" perspective. This approach was particularly crucial for collecting data on two networks that illustrate the differences between a top-down and bottom-up approach to organizing.

The Top-Down Approach

Driving through eastern Iowa one comes to understand the stylized landscapes of Grant Wood's paintings. The hills are abruptly rounded, lined with corn in long, straight rows just as Wood depicted them sixty years ago. If that part of the landscape resembles a Grant Wood painting little else about rural Iowa does. Shiny metal grain bins, ranch houses, and trailer houses frequently replace the red barns and square frame houses. Soybeans now share equal acreage with corn. Hog production has shifted from clover pastures to intensive confined feeding operations on many of the remaining hog farms. In February 1995 I traveled through eastern Iowa to Millersville, to meet with Steve Longley a farmer and organizer of a network (all names and locations are fictitious).

Steve, in his late thirties, recently took over his family's farming operation from his father with whom he had worked since graduating from the agricultural college. With his wife, Janice, he raised pigs from a herd of three hundred sows in a variety of confined and semiconfined buildings. Janice managed the confinement farrowing house and nursery where pigs are born and weaned, while Steve handled the outdoor finishing lots. In addition, Janice kept track of all of the production and financial records on the farm's personal computer. The rate of gain for pigs fattened on outdoor lots suffered during the recent frigid Iowa winters. Steve's new breeding stock produced the lean, finished animals that the packers paid premiums for, but the lean animals needed the warmth of an enclosed finishing unit, otherwise they spent too much energy keeping warm instead of growing.

To solve his dilemma, Steve began talking with nearby farmers about the possibility of forming a network to build new nursery and finishing facilities. After a few informal meetings Steve decided to seek outside planning help. The overall plan came from a group of advisers, Team Pork, organized by Iowa State University Extension Service. Steve and his partners visited the Team Pork advisers in Ames and returned with a *Community Nursery Handbook* to help them develop their plans. Like the NPPC teleconference, the handbook offered a sort of top-down cookie cutter approach. Additionally, to help with plans to create a limited liability corporation, Steve and his partners employed a business planner from Land O' Lakes, a regional cooperative.

All members of the group except two were clustered around the nearby town of Millersville. Bill Craig, Jack Smith, and Al Jones had grown up in the community. Al's partner, Brian Williams, was a transplant from nearby Kelton. The other two members, Bruce Weber and Herman Mueller, lived near Alton, thirty miles to the north. They had heard of the Millersville network plans through their veterinarian and started coming to the planning meetings after a network in their area failed.

The plan that Steve and the other farmers worked on all winter was to build new nursery and finishing facilities. Weaned pigs from each member farm would go to these facilities until they were marketed. The plan was to utilize the relatively new technology called "segregated early weaning" (SEW). Instead of weaning pigs from their mothers at the typical age of six weeks the pigs are weaned at less than three weeks and transferred to "hot" nurseries. These specially built nurseries provide warmth, food, and water for the young pig until it is transferred to a finishing unit for fattening.

The pretwenty-one day weaning is crucial since this is when natural immunities from the mother are actively protecting the piglet from disease. By removing the young pig from contact with older pigs prior to the fading of this natural immunity, exposure to disease is eliminated. This provides several benefits to producers. First, pigs are healthier, thus reducing the costs of medication, veterinary expenses, and death. Second, the rate of gain is improved because of better health. Third, the immunity allows the cohabitation of pigs from different herds.

This new technique poses challenges. Isolation is crucial to SEW success. Baby pigs need to be moved to nurseries that are not just in different buildings, but on sites widely separated from sows or older finishing pigs. The problem is that most hog farmers traditionally located their facilities on a common site, sometimes even in one building where pigs move from one end of the building to the other in an assembly line fashion. The SEW technology has been adopted rapidly by large corporate producers who can construct new operations on isolated plots of land with needed separation distances. Most family farmers do not have that luxury.

The dilemma Steve and his partners faced in financing their project was that a great deal of capital and managerial skills were needed to put the new facility together. In order to gain the efficiencies needed to utilize SEW, these farmers would be forced to abandon part of the production process traditionally performed on their own farms and become farrowing specialists. The project would turn the members into investors in the limited liability company that would own the facilities.

I sat in on an afternoon meeting in the basement of the local Lutheran church while Steve and his partners discussed various aspects of their project—the construction of a large hog production facility to finish pigs raised by the members. Financing was one of the chief issues, and Steve reminded the group that they all needed to submit their financial statements to the group's accountant for inclusion in loan applications. Rather than share their private financial conditions with each other, the members agreed to provide the information to one of the outside advisors they were using. As I listened to this group of relatively young progressive farmers I was impressed with the calm, polite, argument-free manner in which they discussed their plans.

A few weeks later I once again sat in the Lutheran church listening to Steve and his partners discuss their plans. By then, I had spent time getting to know members individually. This night the basement was occupied by a church council meeting, so the network meeting moved to the sanctuary where the farmers sat uncomfortably on hard wooden pews. At one point the pastor looked in and joked about the juxtaposition of God and mammon.

The conversation was much the same as before, but tonight one member, Herman Mueller, was missing. I had not interviewed Herman yet, nor Bruce Weber, another farmer. At the meeting, Bruce sat near the back of the sanctuary and had little to say.

By this meeting, I had learned that one of the members, Jack Smith, would contribute over half of the pigs to the proposed nursery and finishers. In his Ralph Lauren shirt, Jack, in his early forties, looked more like a suburban stock broker than an Iowa hog farmer. He produced pigs from over 1,200 sows, many of which other farmers tended on contract for Jack's family farm corporation. Jack later told me that he often felt that he was "helping out" these farmers who had suffered financial setbacks in hog farming and could no longer obtain credit but who were, nonetheless, skilled operators. The son of a hardware store owner, Jack turned to farming after finishing high school and ran his operation from a modern office at the main farmstead. At the meeting, Jack's input was attentively listened to by other members. Still, each member's opinion carried equal weight despite their differing financial stakes.

As the evening progressed the conversation moved from topic to topic with little formal agenda. At times, the farmers launched into detailed discussions of the merits of building materials and genetic characteristics of swine. The meeting did not end until well after midnight, but the farmers had agreed on answers to all the major questions about building design, scale, and financing. All that remained was to make a loan application. After six months of church basement meetings, the network appeared ready to begin.

What happened over the next few weeks came as a surprise to almost everyone. The first indication of trouble came when I finally interviewed Herman Mueller. Herman was in his midfifties and ran a farm with his wife and son. The Mueller farm, like the Smith farm, was immaculate. The ditches along the road were neatly mowed for a half mile in either direction from the farmhouse. Herman, a Vietnam veteran, was dressed in traditional bib overalls as we sat with his wife, Alice, in the kitchen to talk. Over the phone he had told me he was planning to leave the network. Herman was a straightforward, common sense operator who made it clear that he had no intention of signing a $3 million loan. His reasons had nothing to do with cooperating with other farmers to run the venture. Instead, Herman observed that he would be equal partners with the others, but they would stand to gain more.

"To their credit, they've heard the wake up call," he said, but he added that it was a call he had heard a long time ago. Herman characterized Bill and Steve, both of whom still finished their hogs on open lots, as operating in "caveman fashion." Herman had driven by both farms and considered their generally poor appearance to reflect the farmers who ran them. His own farm was an efficient confinement operation with little debt. Herman wanted to make it more efficient by converting his nursery and finishing units to more farrowing space and placing pigs in an offsite SEW nursery. Bill and Steve, Herman observed, had not made the same needed improvements and were now faced with the need to catch up. While the network would do that for them, it caused Herman to ask why he should be assuming equal risk for far less gain. The fact that they had not modernized earlier caused him to question their skill as managers. It was not so much that he did not want to be in a network. He just did not see how his farm fit with some of the others—their needs were not the same.

That same afternoon I drove down the road to Bruce Weber's farm and heard much the same story. Bruce was young and ambitious, but like Herman he shared a suspicion about acquiring a large debt and working with farmers whose needs differed from his own. Also, like Herman, he did not harbor ill will toward the Millersville farmers, but he noted that whenever a meeting was held a number of issues had been decided beforehand. It was not that the Millersville group was trying push its own agenda, Bruce explained, it was just that the members frequently saw each other at the church and the coffee shop, resulting in conversations and decisions in which he could not participate.

Within a few days, Herman and Bruce officially notified Steve Longley that they were leaving the Millersville network. Steve was disappointed but accepted their decisions without rancor and hoped to recruit other farmers to take their place. With Bruce and Herman out of the picture and the financing ready to go, the debt on the remaining farmers would be larger if they did not scale it back. About this time Al Jones and Brian Williams announced that their participation was unlikely as well. That left Steve and Bill, each with about 200 sows and Jack Smith with his 1,200. To continue, the network would have to alter its plans.

Network planning was put on hold until planting season was over. By early summer, Jack Smith announced that he too had decided against participating. Like the others, he said that the needs of his farm did not match with those of Steve's and Bill's farms. A problem raised by all members, but especially Jack, was swine genetics. Each farmer obtained breeding stock from a different source, and each was loyal to a familiar brand. The process of commingling pigs was, among other things, designed to produce large numbers of relatively uniform animals in hopes of attracting higher bids from meat packers. The lack of common genetics would make that problematic. It would also raise problems with feeding because different genetic lines created variation in rates of gain. Jack's investment in breeding stock made him particularly unwilling to switch to the genetics preferred by Steve and Bill. It was on this point that the one member, one vote rule was stretched too far for Jack. "Why, if I have the most animals, if I contribute the most to the network, should I have to abide by the decisions of the people with so much less at stake?" Jack asked. Jack's departure effectively ended Steve's networking project.

One summer night I sat with Steve and Bill at the dining table in Bill's house as they discussed their options. The formal network planning was now largely useless, and the two men discussed the possibilities for building a much smaller system on one of their own farms. As they tried to figure out a new arrangement, I began to see the flaw that brought down their plans. Reviewing one alternative after another, it became clear that Bill and Steve knew very little about each other's farms. From the type of facilities and their age to feeding practices and ingredients, both men were running very different operations.

Lacking such knowledge, they substituted the top-down approach of the Team Pork plan instead of devising a system responsive to their individual needs. I was surprised that such details had not been covered months earlier until I remembered Steve's early description of how he selected farmers for the group: "They were all progressive enough to accept the networking concept." It was not a matter of selecting farmers because they had common needs, but a matter of fitting Steve's image of progressive farmers, the same image that planners from the agricultural colleges hold.

Weeks later, still no closer to starting a network, Steve blamed the network's failure on the independent nature of farmers. He characterized the farmers he tried to put together as similar to artists, each pursuing his own specialized vision of how a hog farm should be run. What really happened was that the organizer and members of the Millersville network did not know enough about one another at the outset of their venture to evaluate adequately the plan foisted on them. Once they began to assess one another and their own needs, they found the network plan to be a poor fit. Comparative data from a second network illustrates why this happened.

THE BOTTOM-UP APPROACH

Veterinarians, like other small town business operators, have a great deal riding on the future of family farmers. The equity in a veterinary practice is generally considered "blue sky," the anticipated future revenue from a solid client base. Many vets base their retirement plans on the sale of that blue sky to a younger partner. With declining numbers of family hog farms, many vets worry about who will buy their practice down the road. This motivated Allan Garrison to form a swine producer network. In addition, he saw networks as a way of preserving the small town in which he and his wife, Kris, planned to bring up their children.

Located near the Iowa-Illinois border in Kelton, Iowa, the Kelton Pork network was already up and running when I first met Allan Garrison. Allan was in his late thirties, a partner in a six-member veterinary practice that covered several counties in eastern Iowa and northern Illinois. After attending an NPPC forum on networks, he decided to try organizing a network among his customers. As he traveled around the countryside, he took note of the various needs of individual customers. Building on this knowledge, Allan selected a group of farmers who needed to reduce the fat in their hogs to meet the new standards for leanness set by the meat packers.

A dinner meeting in the spring of 1995 marked the network's first anniversary. I sat with a group of farmers as we waited for dinner. Short stretches of conversation were punctuated by awkward silences. I finally asked if everyone at the table knew one another since the silence was uncharacteristic of a group of farmers. "No," came the amused reply from a young farmer sitting next to me, "but we're working on it." The network was formed with farmers from all parts of Allan's trade area, and very few of them were acquainted. Unlike Steve Longley's failed network, which was made up mostly of neighbors who knew each other, here was a successful network of strangers.

As the evening progressed, I learned that Allan had approached each of the farmers with the same pitch, getting together in order to meet a need. Rather than picking a group of farmers who seemed progressive or personally similar, Allan's strategy was to pick from two groups of farmers with complementary needs. The first group consisted of twelve farmers who needed to upgrade the quality of their product for packers. These were younger, mostly small to medium-sized farrow to finish producers relying on family labor to run their farms. Although they needed new breeding stock, the cost of new breeding animals was too high compared to the traditional method of replacing breeding animals from their own herds. The other group was smaller, made up of three feeder pig producers with very small operations. These farmers had off-farm jobs or businesses and farmed only part time. This group faced a dwindling demand for the small groups of pigs they traditionally sold through weekly public auctions at country sale barns. They faced lower prices due to the small numbers of pigs they marketed.

Allan's solution was to arrange for the smaller producers to buy breeding stock from a genetics company and produce replacement gilts to be sold as feeder pigs to the group of farrow to finish producers. This provided a ready market at a premium price for the feeder pig producers and a source of improved genetic stock requiring only a small cash outlay for the farrow to finish producers. The genetics supplier, a small family run company, was willing to go along with the plan because of the volume. Allan described the network as a win-win situation for everyone. He was correct.

As I traveled from farm to farm with Allan, I came to understand how he knew the needs of his clients. Although individual farms varied enormously, their needs were not unique. As I listened to Allan discuss network arrangements with farmers it became clear that any belief in the virtues of independence was not impinging on their desire to be part of the network. They were making tough choices.

Allan's network also had its share of defectors. But Allan did two things to ensure such defections did not destroy the network. First, all of the farmers who were buying gilts had roughly the same size farms. Thus the network was not overly dependent on any one member. Second, the network not only had a large number of members, but it also had a number of farmers waiting to join. I interviewed each of the farmers who defected, and each told the same tale—the new breeding stock had not performed to their expectations. Allan admitted there were problems that he and the breeder were working to eliminate.

The members left for economic reasons, not because the network threatened their independence.

There are both social and economic dimensions at play. In Allan's network, the two dimensions complement each other, while in Steve Longley's network they worked at cross purposes. Allan had a plan with a clear economic benefit to all participants coupled with the social and cultural values that accorded him a trusted position as an organizer. Also, Allan did not benefit directly as part of the network business arrangement. This further enhanced his credibility. Steve Longley attempted to link a group of farmers primarily based on social relationships rather than on economic need. As a farmer, Steve chose from a range of social acquaintances and had limited economic information about them. In fact, all of the farmers I met with knew very little about each other's farms. As one farmer put it, "I don't know how my best friend raises hogs." In contrast, Allan Garrison has a firm economic and operational familiarity with a variety of farms that is important to his ability to provide treatment and advice. The physical separation of individual farmsteads and the lack of communal economic activity, such as the threshing and baling crews of an earlier generation, produce farmers who are unfamiliar with each other's farms. It is this economic isolation that renders sharing of economic information and cooperation difficult between farmers. Instead, intermediaries such as Allan Garrison provide a link through identifying economic need while sustaining network cooperation through traditional rural values of trust and reciprocal need.

This comparison of two networks brings out two points about the urbanization of rural life. On the one hand, the successful network is built on primarily economic grounds—it brings strangers together on the basis of economic need. On the other hand, it is not governed by the complex rules and formal relationships that the agricultural colleges and Team Pork developed. The network that failed was long on the lawyerly language and formal relations recommended from land grant institutions, but it did not work because it failed to meet anyone's economic needs even though the participants were neighbors and knew each other. While the workable network has an economic fit, it is maintained by trust and custom—a social premise—rather than any laws or formal business contracts.

These examples show that people who need each other can get to know each other, but people who know each other may not be able to cooperate if the cooperation is based solely on formal contractual relationships. That is the difference between "top-down" organization engineered by business planners and their lawyers and the "bottom-up" organization based on mutual economic need. The "bottom-up" approach incorporates rural social norms of mutuality and reciprocity in addition to meeting economic needs, while the "top-down" approach may fail to meet economic needs because it lacks appropriate means for keeping people together for common action.

MAKING ALTERNATIVES WORK

Allan Garrison contacted all of the same advisers that Steve Longley did. Rather than accept the top-down approach promoted by agricultural colleges, Team Pork, and others, Allan worked from the bottom up. As I traveled across Iowa interviewing organizers and members of other networks, a pattern emerged. Successful networks were organized by persons who knew a wide variety of farmers and knew the economic dimensions of their farms. They were also tailored to the needs of the members rather than to an idealized plan. Additionally, small networks such as Allan Garrison's reported little community opposition such as that provoked by large-scale corporate facilities. The social compatibility with the local community is as crucial as economic compatibility within the network itself.

Networks represent a refashioning of the old populist goals of cooperatives. Like the early cooperatives, networks are small and responsive to the needs of their members. At the same time they can provide economic benefits not available to individuals. The farmers in Allan Garrison's network feel they have made choices that will enable them to stay in swine production. Networks hold the promise of a flexible, adaptive response on the part of family farmers that can allow them to compete with those interests who want to continue the transformation of American agriculture and rural life along the industrial model. Those who hope to counter this transformation should encourage the formation of networks that respond to the needs of their members and not to the narrow vision of industry planners. Rather than beaming advice to farmers over satellites, those planners might do well to listen to the voices coming from the countryside that offer an alternative future.

REFERENCES

Barnes, J. A. 1954. Class and Committee in a Norwegian Island Parish. *Human Relations* 7: 39–58.

Des Moines Register. 1994. Letters to the Editor. 29 May.

Hightower, Jim. 1978[1972]. *Hard tomatoes, hard times.* Cambridge, Mass.: Schenkman Publishing Co.

Kravitz, Linda. 1974. *Who's minding the co-op? A report on farmer control of farmer cooperatives.* Washington, D.C.: Agribusiness Accountability Project.

Paulsen, Arnold, and Michael Rahm. 1979. *Development of subsidiary sow-farrowing firms in Iowa.* Ames, Iowa: Agriculture and Home Economics Experiment Station.

Rasmussen, Wayne D. 1991. *Farmers, cooperatives, and USDA: A history of agricultural cooperative service.* Washington, D.C.: USDA.

Conclusion
The Urbanization of Rural America

Walter Goldschmidt

The Chapters in this book show what has been happening to rural life in Iowa and elsewhere in rural America. Iowa is the quintessential locale of the American farm, the very stuff of the mythic America, where motherhood and apple pie are not merely the living symbol but, one almost believes, were actually invented. They tell you in scholarly and sober prose what is happening on this rich land, so recently frontier. But if you want to *feel* the changes that have happened to the people living in the towns on the prairies, I recommend that you read one after the other, as I did recently, Willa Cather's *My Ántonia* and Jane Smiley's *Thousand Acres.* Zebulon County, Iowa, is but a few hundred miles and less than one hundred years away from Black Hawk, Nebraska. It is light-years distant in social time.

Cather's Nebraska was just emerging from its sod-house, Indian-fighting existence, and raw new communities were taking form out of the ethnic and social mixture of the frontier. It is not an ideal small-town world, Cather knew that infidelity, jealousy, murder, and all the other tragic dramas of human existence were to be found there. But she also showed that within the ambit of these human foibles there is something else. We see the people of Black Hawk hammering with their incessant labor on the anvil of a generous but demanding soil to forge the American character. Here is the home of those homely virtues: the sense of duty and purpose, the unsentimental sense of community despite ethnic diversity and class difference. Family and farmstead—so intertwined as to be one—are at the center of this new-minted America.

Smiley's Iowa is in the pioneering stages of a new era, the era of industrialized farming. The great grandchildren of Black Hawk's contemporaries are being sold giant, air-conditioned combines by urbane salesmen to work their overexpanded acreage while more remote influences induce them to build the cement-floored apartment suites for their venture into hog production factories while bankers and accountants sell them on incorporation as a way to fend off the predations of the tax system. Just as Cather's Black Hawk saw the dawning of the commercial family farming that shaped the American character, so too Smiley's Zebulon County is at the dawning of industrialized agriculture in the American heartland. Family and farmstead are falling apart. What kind of America will be forged in these new factories?

About half-way through the century that separates these two sets of pioneers, another novel gave us another view of the rural landscape in America: John Steinbeck's *Grapes of Wrath* This work was certainly in my mind as, in 1940, I undertook a study of the social life that industrialized farming creates. My first research was an ethnography of the town of Wasco in California's Central Valley, selected as being an ordinary ("typical") town where field crops were grown on "factories in the field," as Carey McWilliams put it in the title of his book about California agriculture. This was my doctoral thesis that I later published in a book, *As You Sow.*[1] The theme of that book is that from industrialized farming we reap an urbanized kind of rural life: as you sow, so shall you reap.

Later, I made a second study that compared two towns in the same region, Arvin, dominated by giant corporate enterprises, and Dinuba, surrounded by family-sized farms. The evidence clearly demonstrated that the social problems inherent in industrial agriculture are exacerbated by such corporate dominance. The study raised such hackles among departmental bigwigs that it was suppressed by a handful of them, including the secretary of agriculture, until the Senate Small Business Committee learned about it and published it as one of their prints. This study had been made before *As You Sow* was published and some of its conclusions are in that book. Some thirty years later these books were re-issued together under the expanded title, *As You Sow: Three Studies in the Social Consequences of Agribusiness.* The third study promised in the title deals with the propaganda attack on my earlier research, of which more later. In *As You Sow* I prophesied that, failing the adoption of certain recommended policies with respect to taxes, welfare, and the treatment of labor, industrialized farming would spread throughout the nation and that traditional American rural life would become urbanized. In this chapter I want to spell out what this means for the people of Iowa itself, for Iowa, for America.

What is industrialized agriculture? Industrialized farming is large-scale operations with state-of-the-art technology, fully integrated into the market system, dependent upon wage labor under a hierarchical scheme of management. It is efficiency driven and unsentimentally profit oriented. That is to say, management, whether the owner of a local enterprise that has grown beyond the normal operations of a family farm or the CEO of a corporation, makes decisions solely in the interest of profit. Profit means profit for the enterprise and is not to be confused with cost effectiveness, for it takes no account of hidden costs that ultimately may be borne by the community. Research-based standards of efficiency replace tradition in setting farm practices. Industrial operation stands in contrast to the peasant farming of our European forebears where families lived in the village for generations, the peasants subsisting largely on their own produce and having little dependence on the market. The commercial farming that grew rapidly as westering pioneers brought the American heartland under the plow lies somewhere between these extremes, with growing dependence on the market but retaining the close association between family and the farm enterprise and retaining strong ties to the locality.

What is urbanization? It is the way of life of the cities. Cities are large, heterogeneous agglomerates of people where order is maintained by law rather than custom and where human interaction is economic rather than social. Just as the industrial farm enterprise looks outward to the city and national markets, so too the social life looks outward to the centers of power so that individual behavior is shaped by impersonal media-driven city standards. Reflecting the profit orientation of the economic enterprise, social standing is based on monetary wealth. This erodes the sense of community, the ideals of mutuality, and the social value of civility. People from outside control the economy and thus have increasing power over community actions and as a result invade the social culture of the towns and small cities of the countryside.

These industrializing forces had shaped the rural life in California long before my study was made. The Spanish past and the flood of cheap Oriental labor, the Mediterranean climate supporting valuable specialty crops, and the national markets that were opened by the railroads had made California the proving ground for the large-scale, labor-intensive, technologically innovative production as early as the 1880s. They made industrial farming possible but not necessary; social policies were needed for that.

Here are some of the things that I discovered about the social life of the town of Wasco and the other communities I studied that led me to call it urbanized. The towns of first contact for the farmer—the local town—were larger and further apart than in traditional farm communities (twenty or so miles and several thousand people compared to four miles and several hundred people in Iowa at the time). The community was filled with outsiders rather than local farmers and store keepers. The banker, the merchants, the school teachers, and, above all, the laborers were not local people taking on these tasks, but people assigned to the town by the corporation, graduates of colleges taking their first temporary posts, or migrant laborers looking for work. In California, these laborers traditionally had been foreigners of different "race," but at the time of my research during the Depression most of these "foreigners" were blond-haired and blue-eyed Okies. Still, they were viewed with xenophobic, racist hostility.

Those farmers who continued to work their own land and remain in the countryside found that the towns were run by these outsiders, by those sent to the community from the outside. The banker in Wasco was called the "unofficial mayor" of this then unincorporated town. But that "mayor" owed his primary allegiance to the Bank of America, from which he drew his salary and took his orders. This was true of most of those outsiders (whom, ironically, and I now think mistakenly, I called "the insider group" because they were seen as the establishment, in contrast to the laborers). They exerted their influence on civic matters through institutions, also brought in from outside, like service clubs such as Rotary and Twenty-Thirty. Only the most successful farmers (that is, those who had fully entered into the industrialization of their enterprise) took part in such affairs. The working farmers, too busy in their fields, said disdainfully that the avant garde among them were "farming from the bars." This was both right and wrong; the fact is that following the market was far more profitable in this kind of agriculture than following the plow. Plowing was work for hired labor.

The workers lived in the sizable slums that are found in all the towns in the California heartland, the unskilled laborers who truly were the "outsiders" to these towns of outsiders. In Wasco the African Americans were segregated into a ten-acre ghetto, the Okies in a poor neighborhood or in dreadful camps provided by the larger industrial producers or, with luck, in the New Deal's government-created camps. The barrier between the community and these workers was nearly unbreachable; they did not even pray together, for the congregations of the several churches reflected the social strata of the town. Only the public schools enforced a tense and touchy interaction that let a few truly gifted students or athletes find their way across the divide.

Social relationships were money-based and social standing was money-driven. Old values of neighborliness, which are so much the spirit of Cather's Black Hawk, had been so deeply buried under the display of personal achievement through expressions of affluence with household decor, car, and the like, that they were all but invisible. The social elite looked away from the community to the urban centers whether or not, as with the banker, that is where their employer was. The most successful of the local farmers had expanded his dairy farm into a commercial dairy with outlets as far away as Los Angeles, where he centered his own social life. In Arvin I was told about a party in the nearby city of Bakersfield that was of such grandeur that "everybody was there. The Japs could have bombed this town and not killed a soul." Nobody, that is, but the invisible laborers, the ones who in this very town a few years earlier had been beaten as dangerous radicals, some to death, for joining the union.

I could see the vicious cycle at work that led each farmer deeper into this system. Young men representing McCormick-Deering and other industrial enterprises, again like the banker, had been sent by the parent company to be both salesmen and "leaders." I was told laughingly how they oversold the farmers on equipment they did not need. (After all, what red-blooded American man can resist the macho power of these machines?) The farmers then had to expand their farm operations beyond the acreage they owned, abetted by the necessary bank loan, and thereby furthering the consolidation of landholding or farm operation into ever bigger units that would demand still more equipment, and so on. Or go into bankruptcy so that their acreage could enter into the inexorable process of consolidation of landholdings. These pressures apply to herbicides and fertilizers and all the other sophisticated paraphernalia of industrial production.

Such observations led me to predict that industrialized farming would replace the family farmer throughout America unless policies were adopted that would stop the process or, more accurately, unless the policies then in force that had created this vicious cycle were rescinded. (I will come back to this matter later.) The chapters in this volume show that my Cassandra-like predictions are coming true in the very heartland of America. I want, in what follows, to reflect on the lessons in these chapters that show how the factory-like production of pork is reshaping the local farmer, the local community, and the national spirit of America.

The editors of this book invited me to visit Iowa in 1995, giving me my first glimpse of this future. I had an immediate sense of déjà vu. Long stretches of unfenced fields cultivated to a monocrop (corn and soybeans, in this case) running for mile upon mile, hardly ever interrupted by farmsteads like those lovingly depicted on Christmas cards. Local villages, rural schools, country church spires had disappeared by the dozens. These were the visual testimony of the statistics of the changes in farm population and farm size that can be drawn from the census. It evoked the past because it was so reminiscent of the west side of California's Central Valley. The only other place I had seen such long stretches of monocropping was in Communist Romania a decade ago, the monocrop being most dramatic because it formed a sea of bright sunflowers beckoning for miles on end, punctuated by occasional dismal multistoried housing projects for the peasants now reduced to workers. What had shocked me at that time was not that it looked like the California I knew, but the fear that it was a preview of the future American rural landscape. I did not like seeing something similar in fabled Iowa.

While I prefer to deal with human events in more human terms, some people like to see what has happened in figures, so I have brought together some that helps us see the general effect of what is happening. In the half century between 1940 and 1990, the number of people in rural Iowa dropped almost a third, from nearly 1.5 million to just over 1 million while the number of farms dropped by over half and the average size of farms doubled, as you can see in table 12.1. There were only about a fifth as many hog farms in 1992 as there had been right after the war, telling us in figures what must have been many a harrowing loss to old residents.

Kendall Thu and I were driven around the Iowa countryside by Blaine Nickles. As you have read in his contribution to this volume, Blaine has lived the transition, has watched his beloved homeland being transformed into something alien. "Here," he would say every quarter mile or so, "there was a homestead; you can still see where the driveway once was." After a while I learned to spot such archeological evidence. "There," he said, "was the school my kids went to." There was not even an old tether-ball post as a reminiscent stela. We did find the decrepit remnants of one of the old towns of first instance that had once served the farmers. For some reason the post office remained and also a service station, but most of the buildings were boarded up. As we drove about, Thu, with camera (the stethoscope of the anthropologist) around his neck, would get out to take pictures of the giant hog facilities being built, of a Stay Out! sign or whatever. Before he snapped the shutter he would ask whether he was still on public property. He did not want to face a lawsuit. Fear and suspicion, the opposites of mutuality and trust, are the first casualties of urbanization.

The chapters in this volume testify to the fact that Blaine is not just being a sentimental old man, but rather a man who sees clearly when he is being betrayed. Laura DeLind's Parma in Michigan, Robert Morgan's Meadow in North Carolina, and Mark Grey's Storm Lake in Iowa recount the same dismal story. It is the story of rural tranquility being torn apart by the rapacity of out-

Table 12.1
*Some Data on Iowa Farmers and Iowa Hogs**

Year	*Number of Farms*	*Average Farm Size (in acres)*	*Number of Hog Farms*	*Number of Hogs on Farms (in thousands)*
1945	209,000	165	———	———
1954	192,993	177	151,508	13,284
1964	154,162	219	106,184	13,674
1974	126,104	262	59,582	11,477
1982	115,413	283	45,768	14,333
1992	96,543	325	31,790	14,153

*I am indebted to Julia K. Venzke for compiling this data from the *1992 Census of Agriculture* and the *1990 Agricultural Statistics*.

siders whose singular goal is to gain wealth, without the least concern for the welfare of those whose lives their actions are destroying. In all these dolorous tales, the most devastating theme is that once these local communities have been shattered they, like humpty-dumpty, cannot be put together again. Trust destroyed is not easily regained; friendships broken are difficult to mend; the invisible lines that separate factions are indelible.

One of the first visits of the day was at the farm of John McNutt and his wife, Dr. Ilene Lande. There I saw the past meet the future. A third, or was it fourth, generation owner of the land, McNutt and his family represented everything I had imagined about the Iowa social landscape, except made more modern by a professional wife who worked closely with her husband. A thoughtful man with a capable collaborative partner and a bright young son grounded that day, Norman Rockwell-like, because of a black eye from some escapade, were trying hard to maintain an operation that retained the character of a family farm and still be modern. McNutt had put in a modern hog production unit handling some 200 sows that produced an estimated 3,800 piglets per year. I had been taken there so I could see this newfangled operation firsthand, and I was duly impressed—also appalled. (When I finally got to bed late that night, after some seminars and a banquet, for which there had been no time to bathe, I was dismayed to discover that my body was still impregnated with the odor of that morning's visit, though I had not touched an animal in this spotless facility.) Here was the quintessential American farm family, working to live the American dream of progress, taking time out to entertain a visitor for a pleasant family lunch in the warmth of a spacious kitchen. Later, after we had been served beef stew, I realized that they had gone out of their way not to embarrass what might have been a Jewish guest. I was again touched. Their operation is a compromise between the traditional Iowa farm and the pork factories down the road. Can this compromise survive?

Can the McNutts preserve their sense of unity with their dwindling supply of neighbors and still play along with the big operators who dominate the hog production industry? It will be tough, for reasons that are out of their very capable hands. He is dealing with ruthless and powerful operators. As John Ikerd says in his chapter, it will take ingenuity and creativity to survive in a market that has been restructured to the demands of the large, vertically integrated hog producers. You can see the struggle to keep this sense of community in the face of external demands on time and energy in the autobiographical account by Jim Braun, whose very efforts to save traditional values are drawing him into the urban maelstrom.

Just how ruthless and powerful these operators are is made manifest by the detailed lies and misrepresentations made at Parma that Laura DeLind describes. She has shown their total unconcern with the welfare of the local people, making no attempt to use local facilities, hire local people, return anything to the community other than the liquefied excrement of the animals. This is all too literally a shit-on-you attitude. Even the ultimate legal victory was a defeat; the sense of community in Parma was in ruins, the moral version of a bombed-out city after a devastating war.

It would be nice to believe that these events in Michigan were unique. Clearly, they are not. Robert Morgan details for us the manner in which this pattern of pork production invaded North Carolina and destroyed the tranquility of towns such as Meadows. A leading hog producer who was also a state senator quietly inserted legislation into the laws of the state that made it impossible to stop the onslaught of this new production pattern. Some hope was, ironically, later instilled by a massive natural disaster—by an act of God. Meanwhile, the social destruction of the Carolinian countryside reflects what has happened in Michigan. "Things have never been the same" in Storm Lake, Iowa, either since pork packing was brought there in 1982, as Mark Gray recounts. Lies, deceit, and double dealing by the promoters in the service of their own personal gain led to the destruction of trust and mutuality that ruins the sense of community. The results are classic expressions of urbanization: internal conflict; high crime rates; and stress on the schools, medical facilities, and institutions of law enforcement.

I had found that the control of the towns in California lay in San Francisco and Los Angeles, in their financial institutions, their manufacturing and supply systems, their marketing controls, and their propaganda machines. Jim Braun puts his finger on this with an anthropological flair when he shows how two expressions have changed their meaning: *access to capital* and *relationship.* They no longer mean seeing your local banker and getting on with neighbors, they are about having influence in the cities. Morgan tells the same story with a more political idiom, as befits his involvement with North Carolina politics, when he shows how state legislators and corporate lawyers determined what happened in rural areas. And Nickles puts it in stark economic terms when he shows how DeCoster invaded Iowa with his Maine-based corporate money. In independent hog farms, as Ikerd puts it, "the equity and management are on-farm" while in contract production, "most of the capital and management comes from the outside."

These influences are not merely political and economic; they are also propagandistic. I know all too well about that. It was turned on me while I was still engaged in my comparative study, with the Associated Farmers' daily radio hour broadcasting such inaccurate and scurrilous remarks that the station manager made them give me air time to respond. I have detailed this attack on my character and veracity in the re-issue of *As You Sow.* First, during my research, again, when I announced my preliminary results, and a third time, when the report was published two years later, they made every effort to sabotage, suppress, and discredit my work. Political pressure was put on the Department of Agriculture to get me out of the field, and this prevented the planned second phase of the research that was to record data on the quality of life in each of the towns in the area and correlate this with farm size. Having prevented me from expanding the scope of the study, they argued that a comparison of two towns was inadequate. We see similar treatment of the research reflected in the essays in this volume. Both Morgan and Nickles tell us that money pressure on the professors shapes their research to the wishes of the industry. Braun and Nickles both find the dean of agriculture in Iowa to be committed to industrializing Iowa pork production. This is the kind of issue Jane Smiley parodies in *Moo.* Parody or not, it represents reality as Thu and Durrenberger found when industry representatives and officials from land grant universities tried to intimidate them personally and through their university administration. Only the University of Iowa's dedication to the principles of academic freedom provided a safe arena for the continuation of their research and the public dissemination of their findings.

Such propagandistic defamation can be made of cultural and social research because the very nature of our subject makes experimentation impossible and therefore leaves us without hard proof, despite overwhelming evidence. Yet social values cannot be brushed aside, for they have important effects on the mundane aspects of everyday existence. This is clearly shown by the effects on physical health recorded by Donham, on mental health shown by Schiffman and her associates, and on water quality described by Jackson—all powerful reminders that human values have visible life-enhancing involvements. We disregard them to the peril of the whole population. Remember the powerful influence of Rachel Carson's *Silent Spring,* which was vilified and denigrated upon its appearance, but is now recognized as the wake-up call to one of the most vital issues of our time. The colleges of agriculture in California, Iowa and North Carolina, and I daresay elsewhere, have been co-opted intellectually by the business enterprises that foster industrialized farming, giving scant attention to those very social costs that are most visible in the villages and towns of the countryside where their real constituents live.

We are told that what is happening to American agriculture is inevitable; just the cost of "progress." "Would you want to go back to the old Fordson tractor and the mold-board plow?" Of course not; no more than to go back to the scythe or the neolithic flint-bladed sickle. Technology has progressed throughout human history because the "better mousetrap" has its value. But not every innovation is advantageous. The nuclear bomb and current genetic research have raised ethical questions that remind us that not every piece of new technology should be used. I am not being a Luddite; technology can lead to human betterment; change can be valuable, but mindless change is a fool's paradise.

The *social consequences* of new technologies are not inevitable. The urbanization of rural life is not being caused by machines and chemicals but is the result of laws and policies that favor the industrialization of agriculture, put in place by those who profit from it. It was brought to California by governmental policies with respect to labor laws, taxation rules, farm "relief," and the research programs of the agriculture schools. At the time of my research in California the business leaders were openly fostering these policies through the Associated Farmers of California in patterns of behavior not unlike those described by Morgan for North Carolina. The social results did not just happen; they were brought about by propaganda and political pressure. With a kind of tragic irony, they were sold to the public as being necessary to preserve the family farm though they have the opposite effect. For instance, the argument to exclude farm labor from the National Labor Relations Act was justified by saying that farmhands are really like family members, as was once the case in places such as Iowa. It was most definitely not the case in California where industrialization was made possible by the availability of masses of low-paid workers. These originally were Chinese "coolies" brought to the United States to build the trans-continental railroads and left to fend for themselves when the rails had all been laid. The Chinese went into agriculture but soon left for better opportunities in the cities, and new generations of workers were recruited from economically distressed foreign lands: India, Japan, the Philippines, Mexico, and ultimately the Okies from our own dustbowl.[2] As each immigrant group escaped from the low-paying jobs and poor conditions of these farm-factories to seek their own American dreams, new cohorts of workers were recruited in order to keep wages low and prevent unionization, just as we saw happening in Storm Lake, with all the hidden costs of which no one seems to take notice.

Business management tends to fear unions, but most giant corporations had adjusted to the regulations of the National Labor Relations Board and learned to live comfortably and profitably with unionization during the prosperous early postwar decades. (They happily took advantage of the relaxation of these rules by union-busting administrations of recent years.) The Associated Farmers and other agents of corporate farming in California would have none of that, and unionization was openly (and literally) fought in the California fields in the dark days of the Great Depression. Even small farmers tend to identify themselves as businessmen, an attitude that the land grant universities do much to foster, and therefore readily adopt anti-union prejudice, forgetting that ultimately they profit by making the more labor-dependent large operators pay union wages. People do not always recognize their own best interests.

Tax write-off rules and other tax incentives gave impetus to industrial farming in the prosperous years after World War II, inducing ill-conceived forms of development in agriculture. Giant hog farms are one example. Such tax rules artificially heat up programs of dubious value; they also give politically powerful people a money interest in a farming area in which they have no social interest. Laura DeLind tells us that "many high up political persons were invested in the operation" of the Parma hog hotel program and that this enabled the corporation to engage in highly dubious practices. The hog-producer member of the North Carolina legislature changed state laws to help corporate hog producers by giving them tax breaks, by measures to prevent such operations from being sued as a public nuisance, and by restricting zoning authorities. As in Michigan, people in powerful positions such as county commissioners were heavily invested in these operations, with evident conflicts of interest.

In Iowa, Blaine Nickles and his grassroots colleagues found themselves facing the political elite when they tried to stem the growth of hog hotels. Even so, Mark Grey reports, the president of the Iowa Beef Processors complained, when planning to enter hog production in Storm Lake, "that Iowa is biased against big business and the state must change its hog production policies" to attract packers. The governor and other influential public persons seem to have been listening to these corporate voices rather than to the local people.

Faceless outsiders often do not even know what town houses their investments and yet make life and death decisions for the people who live there. Whether they are knaves or good and devout people, their only interest in Parma or Meadows or Storm Lake is the dollar return on their investments. Meanwhile, either by co-opting or by convincing some of the local people, they pollute the physical environment that creates hazards to health and also pollute the social environment that creates social chaos. I need not belabor the point; the evidence can be found throughout this volume.

The most important lesson to be drawn from this discussion is that such collusion and chicanery must take place in order to bring about industrialized production. This puts the lie to the assertion that the social consequences of technological advances are the necessary "price of progress." There is nothing inevitable about the social consequences of technological improvement; they come from ill-considered policies or illegal actions, or both.

This collusion between the investors and governmental agencies gives special advantage to industrialized production and creates an uneven playing field, placing small and independent operators at a disadvantage that has nothing to do with a farm's production efficiency or the farmers' faulty choices of action. It puts the lie to the inevitability of the evolutionary development of large corporate enterprises. Such taking advantage of one's neighbor is particularly abhorrent in American rural society with its deep tradition of mutuality and equality.

There are other patterns of behavior that further tip the field in favor of corporations. Accounting practices is one. I live in the shadow of the entertainment industry in Los Angeles where there is a form of creative accounting we call "Hollywood bookkeeping." No knowledgable author or star would accept the most generous offer of royalties as a percentage of the "net profits," for it is well known that there never are any, however great the gross profit may be. It is all a matter of allocating costs. As far as I know this is legal, if dishonest and a matter of *caveat emptor.* It is not limited to Hollywood. It obscures real costs and profits in a way that can have devastating effects on contractors who are unwary or unable to avoid such traps. In an industry dominated by vertical control it is always possible to shift costs from one level to another, thus obscuring the basis for payment and the source of profit. Its effects on the family hog producers was made crystal clear by Jim Braun, who learned about it the hard way. This system lets them pay a different price for pork to the controlled producers from the price to independent ones. Blaine Nickles makes us aware that the "trickle down" effect of this differential not only makes the ordinary farm unprofitable, but also cuts land values, increases foreclosures, and reduces local tax revenue. We see how these events work themselves out in Mark Grey's account of Storm Lake. John Morrison points out what has happened in the poultry industry, much further down the road to industrial control than pork production. Farmers' income is kept very low by such accounting methods and in other ways. Furthermore, as John Ikerd points out, the "total returns over direct production costs [on family farms] are available for local retail spending" whereas for "corporate operations, only those costs paid in local wages and salaries are available for local spending." Thus corporate operations also inflict damage on the local business community. Ikerd's demonstration reflects the enigmatic finding in my comparative study of Arvin and Dinuba in California where retail sales were nearly 50 percent more per person in the small farm town than in the large farm town, and total retail sales nearly twice as much, though their total farm income was virtually identical. This shows just the opposite kind of "multiplier effect" from that which is assumed in supply-side economic models.

These practices rest on inequities in power and control. The American farmer traditionally has been squeezed between the suppliers of capital, seeds, fertilizers, and equipment, on the one hand, and the packers and processors of farm products, on the other. This situation has been exacerbated by the increased dependency upon the chemicals and controlled seed, on the supplier side, and the increased importance of packaging and marketing to the ever-growing urban population, on the sales side. Such centralized control has taken over the poultry industry. Processors have maintained a price level for the birds that is just high enough to keep the local producer from going out of business under normal circumstances. Morrison shows us how this control works at the local level in poultry and also that hog contracts are essentially the same. This gives pork producers a very grim view of their future. That is why I said that it is going to be hard for the McNutts to preserve their sense of being independent farmers.

An urbanized rural environment is not merely a difference in customs; it is a difference in culture. It is the change that separates the world of Smiley's Zebulon County from Cather's Black Hawk, a difference made up of a myriad of small changes in artifacts, hopes, expectations, and social relations until the world becomes a different world. This is the way cultural evolution works, bringing about new cultures the way biological evolution forms new species.

The chapters in this book deal with specifics, with what has happened in Parma and Storm Lake, in the state of North Carolina, and in the conditions of poultry production, in physical health and human moods. This is the way science works, narrowing the focus and examining those details wherein the devil is said to reside. But science ultimately is about understanding, about putting these bits together to form a patterned whole. It is not the fruit fly that interests the geneticists, but the process of inheritance; it is not the apple's fall that was important to Newton, but the force of gravity. We must take a moment at the close of this volume to look at what these detailed changes in American agriculture portend at a philosophic level.

Cultures are not merely bundles of artifacts and customs; they are ways of looking at life, at looking at one's fellow man, and, in final analysis, at looking at oneself. The way a people make their living sets the character of their culture. It was no mere figure of speech when I said that the American character was forged in towns such as Black Hawk, because the way we earn our livelihood is the central and dominating element in our daily lives, and farming is at the base of the way we earn our living.

After my studies in American society I turned to this issue in my research in East Africa. I studied tribal peoples who had shifted from living mostly off their cattle herds to become hoe farmers in the centuries before colonial powers over-ran native ways of life. These two economies make very different demands on the people and require different talents and daily behavior. I selected tribes where some parts had kept to the old ways and so could compare existing differences as well as looking at historic changes. I learned that they did not just change the work they perform but altered many other elements in their culture: the way they fought their wars, the way they lived on the land and housed themselves, their legal institutions, family life, patterns of collaboration in work, and the social rewards they cherished.[3] Psychological tests even showed that personality varied. The cattle-keeping people were more out-going and emotionally expressive than the farmers, who were more secretive and suspicious.

Because witchcraft is a sneaky way to express suspicion and hatred, the farmers engaged in more of it than the pastoralists. The lesson is an important one. Cultures are integrated wholes, and a change in one area can send shock waves through all the others; when the change is in the basis of subsistence, the effects can be overwhelming.

We must look at some changes that reverberate through the social life of the people whose towns have been invaded by industrial hog production. The farmer's daily pattern of work has shifted from handling machinery and working the soil to managing an enterprise and watching the market, to being the Iowa counterpart of what in California was derided as "farming from the bars." Either this or worse can be expected if the poultry industry is prophesy: essentially being an employee of an absentee owner, taking orders as specified in a contract rather than being an independent entrepreneur. Do I need to evoke the sociology of bureaucracy to spell out what this means in worker satisfaction? It is the difference between the anxieties of independence and the ignominy of dependency. Work that, for all its hardships and disappointments, gave satisfaction in itself, becomes drudgery.

Consider the rewards that successful performance is expected to bring. The industrial enterprise has only one measure of success: monetary profit. This has come to seem so inevitable that we refer to it as "*the* bottom line," as if there were no other kind of gratification. Everyone needs a competence, needs to eat, and deserves to enjoy reasonable comforts, but beyond these physical demands the focus on money stifles social life. It denies those social rewards that can come only from a sense of belonging to a community and those spiritual satisfactions that derive from a sense of public purpose.

This leads us to the sense of the loss of community about which several of the chapters are quite explicit. Groups are not merely a way to collaborate; they are an essential matrix by which each person gets a sense of self. Humans were created to live in communities and are so constituted that sense of self rests on some kind of group acceptance. At best, the very creation of industrial hog farms pits neighbor against neighbor, because those who profit from introducing them are doing so at the expense of their neighbors—if only by invading their space with foul odors. Trust is a fragile commodity. We have seen the schisms that invade the towns destroy the trust which lies at the base of community life. Cather describes its uncertain beginnings; Smiley tells us that it is lost, even within the families that have substituted monetary for social values. We get insight into the loss of community when we read Randy Ziegenhorn's description of the way farmers try to create them. With modern transportation and communication technology, it is no longer necessary to build social solidarity on the locality, so we see the formation of effective networks of people with similar or complementary interests.

I am not concerned only with the small towns of Iowa or, for that matter, only with the rural United States. The destruction of traditional values in the rural areas is a loss to the culture of America. One might think that the farm sector has become so small in America and so out of the mainstream that it makes little difference to our national scene. I do not think this is the case. The farmer historically has had a special place in our national identity and in shaping our national character, a place earned by virtue of the heroic use of an unusual blend of individual independence and social mutuality. These qualities are built into the way the family farm conducts its daily operations. The successful traditional farmer was both labor and management and a jack-of-all-trades. The farm wife was not a passive helpmate, but a partner in the enterprise. She had no need to prove her independence (which made her impatient with the feminist movement) because she knew she was a vital member of a team. The children apprenticed the many trades that go into the farm operation and bring rich and diverse experience to their adult lives. Each farm family was essentially master of its own fate, responsible for its economic well-being, and yet always intimately tied to a local community. The institutions of the towns and countryside offered a source of mutual aid and a context for those social gratifications which everyone must have. Thus, cold personal independence was played out within the warmth of strong social ties.

I do not want to paint an idyllic picture, I have read too much about the long history of exploiting the farmer, about his vulnerability in the marketplace. I also know that the small towns can be stultifying, filled with pious gossip and harsh judgments. I am not talking about an ideal life—I am not even saying that it is pleasurable—but about choices and preferences, about the character the life builds and therefore about the social model that this mode of life puts before the national consciousness. Perhaps it has even been good that the children on these farms learned the hard work and heartbreaks of farming and the constraints of small towns, so they often were eager to escape its grip, since their migration to the cities has had a leavening effect on our urban life. They have placed a valuable stamp on our national culture. We discard this model at great cost to the nation, a cost that is already beginning to show.

Industrial farming has undermined this tradition by making financial gain the single overriding aim of the productive process. Not only does it destroy communities, but also it deprives the farmers of enjoying the gratifications that farming as a calling can provide. As yet the change has not been fully realized, and it is still possible to stay its hand. Humans, alone among animals, can shape their evolution by perceiving dangers ahead and applying their own wisdom; that is the advantage of cultural over biological evolution. The chapters in this volume should make the people of Iowa and other areas being threatened with similar dangers force those in positions of leadership to formulate policies and enact laws that will prevent the disasters inherent in an unbridled industrialized agriculture.

NOTES

1. My major publications on American agriculture are (1) *As You Sow,* originally published by Harcourt Brace, in 1947 and later by The Free Press; (2) the study sponsored by the Bureau of Agricultural Economics of the U.S. Department of Agriculture, but, because the department was pressured by the Associated Farmers not to release it, was brought out in December 1946 as a special print under the title *Small Business and the Community: A Study of the Effects of Scale of Farm Operations,* as a Report of the Special Committee to Study Problems of American Small Business; U.S. Senate, 79th Congress, 2nd session (Senate Committee Print No 13); and (3) the re-issue of both as *As You Sow: Three Studies in the Social Consequences of Agribusiness,* by Allenheld, Osmun & Co. in 1978. The third study in this book, "Agribusiness and Political Power," is a detailed description of the efforts to suppress the comparative study.

2. The best and most comprehensive statement demonstrating this was made in the doctoral dissertation of Varden Fuller at the University of California, Berkeley, and published by the LaFollette Hearings. See "The Supply of Agricultural Labor as a Factor in the Evolution of Farm Organization in California." *Hearings before a Subcommittee of the Committee on Education and Labor.* United States Senate, 76th Congress, 3rd Session, pursuant to S. Res. 266 (74th Congress). Exhibit 8762A, Part 54. Washington D.C., 1940

3. The major publications dealing with this study are my *Culture and Behavior of the Sebei,* U. California Press 1976 and a shorter, *The Sebei, A Study in Adaptation,* Holt, Rinehart and Winston, 1986. My colleague, Robert Edgerton, dealt in detail with personality attributes in *The Individual in Cultural Adaptation,* 1971, also the University of California Press.

Contributors

Kendall M. Thu is adjunct assistant professor of anthropology, Department of Anthropology, and associate research scientist, at the University of Iowa. He received his Ph.D. in anthropology in 1992 based on ethnographic field research in Norway examining the relationship between state agricultural policy and farming. His two current areas of research are the industrialization of agriculture and rural health in its broadest sense and agricultural change and farm-related injuries among farmers and their families. He has published a number of articles and book chapters on the relationships among state agricultural policies, farm practices, and rural conditions in Norway and the United States. He is the general editor of *Understanding the Impacts of Large Scale Swine Production* and has provided over 50 lectures, presentations, and testimony on swine industry changes to local, state, and national farm, policy, and rural organizations in the past three years.

E. Paul Durrenberger was a professor of anthropology at the University of Iowa from 1972 to 1997. He joined the Department of Anthropology at Pennsylvania State University in 1997. Durrenberger received a Ph.D. in 1971 from the University of Illinois at Champaign-Urbana. He has conducted a wide range of ethnographic field research in Thailand, Iceland, Mississippi, Alabama, Chicago, and Iowa. Much of his work focuses on food production systems and their relationship to human adaptation. Durrenberger has published over one hundred articles and a dozen books on these and related topics, as well as contributing commentary to National Public Radio. He has also served as president of the Society for Economic Anthropology and Culture and Agriculture within the American Anthropological Association.

Laura B. DeLind is a senior academic specialist in the Department of Anthropology at Michigan State University. Much of her research and writing focuses on the structure of industrial U.S. agriculture and the contemporary food system, with particular attention paid to the long-term costs and social and political inequities contained therein, especially as they are felt at the local or community level. DeLind is a proponent of sustainable agriculture and actively works to promote more decentralized, diversified, and democratized systems of food production, distribution, and consumption throughout the Midwest and Michigan.

Jim Braun is a lifelong resident of north-central Iowa. He graduated from CAL Community School in Latimer, Iowa, and studied at Iowa State University. He is a member of the Immanuel Unity Church of Christ, sits on the Franklin County Republican Central Committee, is vice president of Franklin County Pork Producers and president of and lobbyist for Friends of Rural America. His twenty-seven years of independent confinement hog production give him a unique perspective on the changes in the pork industry. **Pam Braun** is a homemaker and artist. She has devoted over twenty-five years to volunteer work in her church, local school, area communities, and through Friends of Rural America. She and Jim live and farm in Latimer with their three children.

Mark A. Grey is associate professor of anthropology at the University of Northern Iowa. He received his Ph.D. in applied anthropology at the University of Colorado-Boulder. His current research interests include rural communities, economic development, and ethnic relations in rapidly changing rural towns in the American Midwest. He also has extensive experience conducting research in rural schools. His recent publications have appeared in *The Annals of Iowa, Sociology of Sport, Research in Social Problems and Public Policy,* and the books *Any Way You Cut It: Meat Processing and Small-Town America and Unionizing the Jungles.*

Kelley J. Donham is professor of preventive medicine and environmental health, College of Medicine, at the University of Iowa. He received his doctorate in Veterinary Medicine from Iowa State University in 1971. He grew up in the Iowa City, Iowa, area on his family's diversified farm that had strong roots in hog production. He currently serves as chairman for the National Coalition for Agricultural Safety and Health and is the current president of the American College of Veterinary Preventive Medicine. Internationally, he serves on the Board of Directors for the International Association for Rural Health and Agricultural Medicine and chairs the Rural and Agricultural Health Committee of the International Commission of Occupational Health. He has published over ninety manuscripts, numerous book chapters, and a book on rural health and agricultural medicine.

Susan Schiffman is professor of medical psychology and director of the weight loss unit in the Department of Psychiatry at Duke University Medical Center in Durham, North Carolina. She is an internationally recognized authority on taste and smell and their role in nutrition and human behavior. Her research spans the range from clinical to molecular investigations of the senses of taste and smell. She has published over two hundred articles, and her research has appeared in many journals, including the *New England Journal of Medicine, Proceedings of the National Academy of Sciences,* and the *Journal of the American Medical Association.* She is the co-editor-in-chief of *Physiology and Behavior* and contributing editor for numerous journals, including *Neuroscience, Biobehavioral Reviews,* and *Primary Sensory Neuron.*

Laura Jackson is assistant professor of biology at the University of Northern Iowa where she teaches courses in conservation biology, applied ecology, agroecology, and environmental studies. She received a bachelor's degree in biology from Grinnell College and a Ph.D. in ecology and evolutionary biology in 1990 from Cornell University. She is currently collaborating with farmers to incorporate native prairie plants into their grazing systems.

Blaine Nickles is a life-long resident of Wright and Hamilton counties in north-central Iowa. He was raised on a family grain and livestock farm and started farming with his wife in 1950. They raised cattle and bred and marketed hogs until 1993. Nickles has been active in Iowa's Organization for the Protection of the Environment (OPE) and a related executive committee that works with various farm commodity groups. In 1994 he was appointed to the Iowa Governor's Environmental Agricultural Task Force.

Robert Morgan is the senior partner in the law firm of Morgan and Reeves in Raleigh, North Carolina. He commenced practice in 1955 and later served two terms as attorney general of North Carolina. His legal practice, private and public, has involved issues of public interest, such as consumer protection, restraint of trade, price fixing, and rate making by utilities and insurance companies. Morgan also served as United States senator from North Carolina from 1975 through 1981.

John M. Morrison is the executive director of the National Contract Poultry Growers Association and has directed the day-to-day operations of this agricultural cooperative since July 1992. The Board of Directors selected him in light of his poultry farming experience and leadership background. In 1988, Morrison left his position as president of a Dallas-based oil and gas exploration company and moved to Louisiana to build a breeder hen farming operation. He contracted with ConAgra Poultry Company to produce hatching eggs until the time his farm was sold in 1992 and he began working with the Contract Poultry Growers Association.

John E. Ikerd is extension professor of agricultural economics and coordinator of Sustainable Agricultural Extension programs, University of Missouri, Columbia, Missouri. He earned his Ph.D. in agricultural economics from the University of Missouri. Ikerd worked three years with Wilson Foods prior to his graduate studies. He held Extension Economics positions at North Carolina State University, Oklahoma State University, and the University of

Georgia before returning to the University of Missouri in 1989 to provide state and national leadership for research and education programs related to sustainable agriculture.

Randy Ziegenhorn is a farmer from New Boston, Illinois. He received a Ph.D. in anthropology from the University of Iowa in 1997 based on ethnographic field research among Iowa pork producers and others involved in production networks. He also has studied networks of cooperation and competition in Iowa's hybrid seed corn industry.

Walter Goldschmidt took his doctorate in anthropology in 1942 from the University of California, Berkeley. His thesis was the study of the community of Wasco, California, which, together with subsequent research for the Bureau of Agricultural Economics of the USDA, laid the basis for his discussion in this book. Goldschmidt also did research on California Indian tribes and on land rights and uses of the Indians of southeast Alaska. After producing a series of radio dramas to popularize anthropological understandings ("The Ways of Mankind," 1951–53) he went to Africa to examine the effects of economic infrastructure on tribal life and culture by studying peoples who had shifted from cattle keeping to hoe farming. He went to UCLA in 1946 where he spent the remainder of his academic career, with secondary appointments in sociology and psychiatry, retiring in 1991. Goldschmidt is also the author of three books in social theory, the most recent of which is *The Human Career.* He served as editor of the *American Anthropologist* (1954–59), was founding editor of *Ethos,* and is past president of the American Anthropological Association and the Society for American Ethnology.

Index

D

E

F

G

H

I

J

K

L

R

S

T

U

V

W